Real Estate Math Express

"Rapid Review and Practice with Essential License Exam Calculations"

Exam Review
Workbook

Second Edition

Performance Programs Company

Stephen Mettling
David Cusic
Ryan Mettling

ISBN: 978-0915777112

Preface

Purpose and Benefits of Real Estate Math Express

Real Estate Math Express is a concise, easy-to-study test preparation guide to help real estate students improve their real estate math scores to pass the state licensing test.

The primary feature of Real Estate Math Express is that it contains all necessary formulas and practice questions in 70+ pages. This enables you to truly cram for the math portion of your state's licensure test, memorize key formulas, and walk into the exam site having all the essential material in your short-term memory! You simply cannot accomplish this using competing lengthy exam prep books.

So, if you want a truly rapid review resource, Real Estate Math Express is for you. No frills, no lengthy explanations.....just key formulas, definitions, corresponding examples and practice questions.

Contents

Real Estate Math Express can be broken down into 3 major parts:

- Math Formulary
- Math Questions
- Answer Key

The math formulary is the section that introduces each major real estate math topic and its corresponding formula(s). Real Estate Math Express covers numerous topics ranging from basic math to capitalization rates to prorations and finally to closing cost calculations.

Once the math formulas have been mastered, there are over 110 sample questions for you to practice with. Each question has both an answer and an explanation of the correct answer at the end of the book.

About the Authors

For over thirty years, Stephen Mettling and David Cusic have operated one of the nation's most successful custom training organizations specializing in real estate program development. They are most known for writing Principles of Real Estate Practice, a national real estate principles textbook used by over 35,000 students per year. Stephen Mettling has also served as vice president and author for a major real estate training and publishing organization. Dr. Cusic, an author and educator, has been engaged in vocation-oriented education since 1966. Specializing in real estate training since 1983, he has developed numerous real estate training programs for corporate and institutional clients around the country including NAR, CoreNet Global, and CCIM.

Ryan Mettling graduated valedictorian from the University of Central Florida's College of Business Administration with a B.S. in Business Management. He is an accomplished online curriculum designer and course developer. Ryan currently serves as the publisher of Performance Programs Company.

Table of Contents

Math Formulary

Basic Math

Fractions

Adding and subtracting same denominator:

Formula: $$\frac{a}{c} + \frac{b}{c} = \frac{a+b}{c}$$

Example: $$\frac{1}{2} + \frac{1}{2} = \frac{1+1}{2} = \frac{2}{2} = 1$$

Adding and subtracting different denominators:

Formula: $$\frac{a}{c} + \frac{b}{d} = \frac{ad+bc}{cd}$$

Example: $$\frac{1}{2} + \frac{1}{3} = \frac{3+2}{6} = \frac{5}{6}$$

Multiplying:

Formula: $$\frac{a}{c} \times \frac{b}{d} = \frac{ab}{cd}$$

Example: $$\frac{2}{5} \times \frac{4}{6} = \frac{8}{30} = \frac{4}{15}$$

Decimals and Percents

Converting decimals to percentages

      Formula: *Decimal number x 100 = Percentage number*

      Example: .022 x 100 = 2.2%

Converting percentages to decimals

      Formula: $\dfrac{Percent\ number}{100} = Decimal\ number$

      Example: $\dfrac{2.2}{100} = .022$

Multiplying percents

      Formula: 1. $\dfrac{Percent\ number}{100} = Decimal\ number$

                     2. *Beginning number x Decimal number = Product*

      Example: 75% of 256 (75% x 256) = ?

                     1. $\dfrac{75}{100} = .75$

                     2. 256 x.75= 192

Dividing by percents

      Formula: 1. $\dfrac{Percent\ number}{100} = Decimal\ number$

                     2. *Beginning number ÷ Decimal = Dividend*

Example: 240 ÷ 75% =   ?

    1. 75% ÷ 100 = .75

    2. 240 ÷ .75 = 320

Decimals, Fractions, and Percentages

Converting fractions to percents

    Formula: (1)   a / b or a ÷ b = a divided by b = decimal number

             (2)   decimal number x 100 = percent number

    Example: (1)   2 / 5 = 2 divided by 5 = 0.4

             (2)   .4 x 100 = 40%

Converting a percent to fraction and reducing it

    Formula: (1)   X% = X ÷ 100 or X / 100

             (2) $\dfrac{X \div a}{100 \div a}$ where a is the largest number that divides evenly into both numerator and denominator

    Example: (1)   40% = 40 ÷ 100, or 40 / 100

             (2) $\dfrac{40 \div 20}{100 \div 20} = \dfrac{2}{5}$

Converting fractions to decimals and percentages

    Formula: *Decimal x 100 = Percent number*

    Example: .75 x 100 = 75%

Equations

Additions and Subtractions

Formula: *if* $a = b + c$

   then $b = a - c$ *(subtracting c from both sides)*

   and $c = a - b$ *(subtracting b from both sides)*

Example: $10 = 6 + 4$

   $6 = 10 - 4$

   $4 = 10 - 6$

Multiplications and Divisions

Formula: *if* $a = b \times c$

   then $b = \dfrac{a}{c}$ *(dividing both sides by c)*

   and $c = \dfrac{a}{b}$ *(dividing both sides by b)*

Example: $10 = 2 \times 5$

   $2 = \dfrac{10}{5}$

   $5 = \dfrac{10}{2}$

Linear and Perimeter Measurement

Linear measure of rectangles

Formula: $Side\ A = \dfrac{Area}{Side\ B}$

Example: A rectangular house has one side 40' side long and area of 1,200 SF. What is the length of the other side?

Side A = (1,200' ÷ 40') = 30'

Perimeter measurement

Formula: *Perimeter = Sum of all sides of an object*

Example: A five-sided lot has the following dimensions:

Side A = 50' Side B = 60'
Side C= 70' Side D = 100'
Side E = 30'

What is the perimeter of the lot?

P = 50' + 60' + 70' + 100' + 30' = 310'

Area Measurement

Square and rectangle

Formula: *Area = Width x Depth (Horizontal) or Width x Height (Vertical)*

Width= Depth (Height) ÷ Area

Depth (Height)= Width ÷ Area

Example: A house is 40' deep and 30' wide. What is its area?

Area = 40' x 30' = 1,200 SF

Triangle

Formula: $Area = Height \times \dfrac{Base(Width)}{2}$

Example: An A-frame house has a front facade measuring 30' across and 20' in height. What is the area of the facade?

Area = (30' x 20') ÷ 2 = 300 Square feet (SF)

Square foot-to-acre conversion

Formula: $Acres = \dfrac{Area\ SF}{43,560\ SF}$

Example: How many acres is 196,020 SF?

196,020 SF ÷ 43,560 SF = 4.5 acres

Acre-to-square foot conversion

Formula: *SF = Number of acres x 43,560 SF*

Example: How many square feet is .75 acres?

.75 acres x 43,560 SF = 32,670 SF

Linear and Area Conversion Chart

Linear measures

(cm = centimeter; m = meter; km = kilometer)

1 inch	=	1/12 foot	=	1/36 yard	
1 foot	=	12 inches	=	1/3 yard	
1 yard	=	36 inches	=	3 feet	
1 rod	=	16.5 feet	=	1/320 mile	
1 mile	=	5280 feet	=	1760 yards	= 320 rods

1 centimeter	=	1/100 m		
1 meter	=	100 cm	=	1/1000 km
1 kilometer	=	1,000 m		

Area measures

1 square inch	=	1/144 sq. foot		
1 square foot	=	1/9 sq. yard	=	144 sq. inches
1 square yard	=	9 sq. feet	=	1,296 sq. inches

1 acre	=	1/640 sq. mi	=	43,560 SF	= 208.71 ft x 208.71 ft
1 square mile	=	640 acres	=	1 section	= 1/36 township
1 section	=	1 mi x 1 mi	=	640 acres	= 1/36 township
1 township	=	6 mi x 6 mi	=	36 sq. mi	= 36 sections

Metric conversions

(cm = centimeter; m = meter; km = kilometer)

1 inch	=	2.54 cm				
1 foot	=	30.48 cm	=	.3048 m		
1 yard	=	91.44 cm	=	.9144 m		
1 mile	=	1609.3 m	=	1.60 km		
1 centimeter	=	.3937 inch				
1 meter	=	39.37 inches	=	3.28 feet	=	1.094 yards
1 kilometer	=	3,281.5 feet	=	.621 mile		

Fractions of sections, acres, and linear dimensions

Fraction		# Acres		Feet X Feet
1 section	=	640 acres	=	5280 X 5280
1/2 section	=	320 acres	=	5280 X 2640
1/4 section	=	160 acres	=	2640 X 2640
1/8 section	=	80 acres	=	2640 X 1320
1/16 section	=	40 acres	=	1320 X 1320
1/32 section	=	20 acres	=	660 X 1320
1/64 section	=	10 acres	=	660 X 660

Calculating Area from the Legal Description

Formula: (1)   *First multiply all the denominators of the fractions in the legal description together.*

        (2)   *Then divide 640 by the resulting product.*

Examples: How many acres are in the Northern 1/2 of the Southwestern 1/4 of Section 6?

$$\frac{640}{2 \times 4} = \frac{640}{8} = 80 \text{ acres}$$

How many acres are in the Western 1/2 of the Northwestern 1/4 of the Northeastern 1/4 of Section 8?

$$\frac{640}{2 \times 4 \times 4} = \frac{640}{32} = 20 \text{ acres}$$

Volume Measurement

Formula: *Volume = Width x Height x Depth*   *(assume objects with 90 degree angles)*

$$Base = \frac{Height \times Depth}{Volume}$$

$$Height = \frac{Base \times Depth}{Volume}$$

$$Depth = \frac{Base \times Height}{Volume}$$

Example: What is the volume of a 40' x 30' x 20' house?

40' x 30' x 20' = 24,000 cubic feet

Ownership

Condominium Assessment Calculation

Formula:

$$\text{Monthly assessment} = \frac{\text{Total annual building budget} \times \text{Condo unit \% of value}}{12}$$

Example: Assume a condominium complex with a $300,000 budget and a unit that comprises 1.48% of the total value of the complex. Monthly assessment?

($300,000 x .0148) ÷ 12 = $370 monthly assessment.

Leases

Percentage Lease Rent Calculation

Formula: *Monthly percentage rent = Sales times % Sales charged*

Example: A store generates $50,000 / month. The lease calls for 1.5% percentage rent. Monthly rent amount?

($50,000 x .015) = $750 / month

Contracts for the Sale of Real Estate

Percentage of Listing Price Calculation

Formula: *Percentage of listing price = Offer ÷ Listing price*

Example: A property listed for $150,000 receives an offer for $120,000. The offer's percentage of listing price is:

$120,000 ÷ $150,000 = 80%

Earnest Money Deposit Calculation

Formula: *Deposit = Offering price x required or market-accepted percentage*

Example: A seller requires a 2% deposit on a property listed for $320,000. The required deposit (assuming a full price offer) is:

$320,000 x 2% = $6,400

Appraisal & Value

Adjusting Comparables

Rules:
1. Never adjust the subject!

2. If the comparable is **superior** to the subject, **subtract** value from the comparable.

3. If the comparable is **inferior** to the subject, **add** value to the comparable.

Example: The subject has a $10,000 pool and no porch. A comparable that sold for $250,000 has a porch ($5,000), an extra bathroom ($6,000), and no pool.

Adjustments to comp: $250,000 (+10,000 - 5,000 - 6,000) = $249,000 indicated value of subject

Gross Rent Multiplier

Formulas: *Sales price = Monthly rental income x GRM*

$$Monthly\ rental\ income = \frac{Sales\ price}{Gross\ Rent\ Multiplier}$$

Examples: 1.   What is the value of a fourplex with monthly rent of $2,800 and a GRM of 112?

$2,800 rent x 112 GRM = $313,600

2.   What is the GRM of a fourplex with monthly rent of $2,800 and a value of $313,600?

$313,600 price ÷ $2,800 rent = 112 GRM

Gross Income Multiplier

Formulas $$Gross\ Income\ Multiplier = \frac{Sales\ price}{Annual\ income}$$

Sales price = Annual income x GIM

$$Annual\ income = \frac{Sales\ price}{Gross\ Income\ Multiplier}$$

Examples: 1.   What is the value of a commercial property with an annual income of $33,600 and a GIM of 9.3?

$33,600 income x 9.3 GIM = $312,480

2.   What is the GIM of a commercial property with annual income of $33,600 and a value of $312,480?

$313,600 price ÷ $33,600 = 9.3 GIM

Cost Approach Formula

Formula: *Value = Land value + (Improvements + Capital additions - Depreciation)*

Example: Land value = $50,000; home replacement cost = $150,000; new garage added @ $30,000; total depreciation = $10,000

Value = $50,000 + (150,000 + 30,000 - 10,000) = $220,000

Depreciation

Formulas:
$$\text{Annual depreciation} = \frac{\text{Beginning depreciable basis}}{\text{Depreciation term (Number of years)}}$$

Depreciable basis = (Initial property value + Any capital improvements - Land value)

Example: Property value = $500,000; land value = $110,000; depreciation term = 39 years

Step 1: ($500,000 - 110,000) = $390,000 depreciable basis

Step 2: ($390,000 ÷ 39 years) = $10,000 annual depreciation

Income Capitalization Formula

Formulas:
$$Value = \frac{\text{Annual Net Operating Income}}{\text{Capitalization rate}}$$

$$\text{Capitalization rate} = \frac{\text{Annual Net Operating Income}}{Value}$$

Annual Net Operating Income = Value x Capitalization rate

Examples: 1. A property generates $490,000 net income and sells at a 7% cap rate. What is its value?

$490,000 ÷ 7% = $7,000,000 value

2. A property has a net income of $490,000 and sells for $7,000,000. What is its cap rate?

$490,000 ÷ 7,000,000 = .07, or 7%

3. A property's value is $7,000,000 and the cap rate is 7%. What is the property's net operating income?

$7,000,000 x .07 = $490,000

Net Operating Income (NOI, Net Income)

Formula: *NOI = Potential rent - Vacancy loss + Other income - Operating expenses*

Note: NOI does not include debt payments!

Example: A building has 10 office suites generating annual potential rent of $10,000 each. Vacancy = 10% and annual expenses are $35,000. Vending machines yield $5,000. What is the NOI?

$100,000 rent - 10,000 vacancy + 5,000 other income - 35,000 expenses = $60,000 NOI

Finance

Points

Definition: 1 point = 1% of the loan amount or .01 x loan amount

Formulas: $Points = \dfrac{Fee\ paid}{Loan\ amount}$

Fee paid = Loan amount x Points

Examples:

1.   A borrower pays $500 for a $10,000 loan. How many points are paid?

   $500 ÷ 10,000 = .05 = 5 points

2.   A borrower pays 5 points on a $10,000 loan. What is the fee paid?

   $10,000 x .05 = $500

3.   A borrower pays $500 as 5 points on a loan. What is the loan amount?

   $500 ÷ .05 = $10,000

Rules of Thumb: 1 point charged raises lender's yield by .125%

8 points charged raises lender's yield by 1%

Example: A lender wants to yield 7% on a 6.5% loan. How many points must he or she charge?

(7% - 6.5%) = .5%

$\dfrac{.5\%}{.125\%} = 4 \text{ points}$

Interest Rate, Principal and Payment

Caveat!   Interest rates in mortgage financing apply to the annual interest payment and exclude principal payment. Remember to convert annual payments to monthly or vice versa as the question requires, and to exclude principal payments from your calculations!

Formulas: *Payment = Principal x Rate*

$$Principal = \frac{Payment}{Rate}$$

$$Rate = \frac{Payment}{Principal}$$

Examples: 1.   A borrower has a $100,000 loan @ 6% interest. What are the annual and monthly payments?

Annual payment = $100,000 x .06 = $6,000
Monthly payment = $6,000 ÷ 12 = $500

2.   A borrower has a $500 monthly payment on a 6% loan. What is the loan principal?

Principal = ($500 x 12) ÷ 6% = ($6,000 ÷ .06) = $100,000

3.   A borrower has a $500 monthly payment on a $100,000 loan. What is the loan rate?

Rate = ($500 x 12) ÷ $100,000 = ($6,000 ÷ 100,000) = .06 = 6%

Total Interest, Interest Rate, and Loan Term

Formulas: *Interest-only loan:*   *Total interest = Loan amount x Rate x Term in years*

Amortized loan: *Total interest = (Monthly PI payment x 12 x term) - Loan amount*

Examples: 1.   A borrower obtains a 10-year interest only loan of $50,000 @ 6%. How much interest will he or she pay?

($50,000 x .06 x 10) = $30,000

2.   A borrower obtains a 10-year amortized loan of $50,000 @ 6% with monthly payments of $555.10. How much interest will he or she pay?

   ($555.10 x 12 x 10) - $50,000 = $16,612

Amortization Calculation

Formulas: *Month 1:*   *Principal paid = Monthly payment - (Loan amount x Rate ÷ 12)*

   Month 2:   *New loan amount = (Previous month principal - Principal paid)*

   Principal paid = Monthly payment - (New loan amount x Rate ÷ 12)

Example: A borrower obtains a 30-year $100,000 amortized loan @ 7% with a $665.31 monthly payment. What is the principal paid in the second month?

   Month 1:   Principal paid = $665.31 - ($100,000 x 7% ÷ 12) = $665.31 - (583.33 interest paid) = $81.98

   Month 2:   New loan amount = $100,000 previous month beginning loan amount - $81.98 principal paid = $99,918.02

   Principal paid = $665.31 - ($99,918.02 x 7% ÷ 12) = $665.31 - (582.86 interest paid) = $82.45

Loan Constants

Formulas: $Monthly\ payment = \dfrac{Loan\ amount}{1000} \times Loan\ constant$

   $Loan\ amount = \dfrac{Monthly\ payment}{Loan\ constant} \times 1000$

   $Loan\ constant = \dfrac{Monthly\ payment}{Loan\ amount} \times 1000$

Examples: 1. A borrower obtains a loan for $100,000 with a 6.3207 constant. What is the monthly payment?

Monthly payment = ($100,000 ÷ 1,000) x 6.3207 = $632.07

2. A borrower has a monthly payment of $632.07 on a loan with a monthly constant of 6.3207. What is the loan amount?

Loan amount = ($632.07 ÷ 6.3207) x 1000 = $100,000

3. A borrower obtains a loan for $100,000 with a monthly payment of $632.07. What is the loan constant?

Loan constant = ($632.07 ÷ $100,000) x 1,000 = 6.3207

Loan - to - Value Ratio (LTV)

Formulas: $$LTV\ ratio = \frac{Loan}{Price\,(Value)}$$

$$Loan = LTV\ ratio \times Price\ (Value)$$

$$Price\ (Value) = \frac{Loan}{LTV\ ratio}$$

Examples: 1. A borrower can get a $265,600 loan on a $332,000 home. What is her LTV ratio?

LTV Ratio = $265,600 ÷ 332,000 = 80%

2. A borrower can get an 80% loan on a $332,000 home. What is the loan amount?

Loan = $332,000 x .80 = $265,600

3. A borrower obtained an 80% loan for $265,600. What was the price of the home?

Price (value) = $265,600 ÷ .80 = $332,000

Financial Qualification

Income ratio qualification

Formula: *Monthly Principal & Interest (PI) payment = Income ratio x Monthly gross income*

Example: A lender uses a 28% income ratio for the PI payment. A borrower grosses $30,000 per year. What monthly PI payment can the borrower afford?

Monthly PI payment = ($30,000 ÷ 12) x .28 = $700

How much can the borrower borrow if the loan constant is 6.3207? (See also- loan constants)

Loan amount = ($700 ÷ 6.3207) x 1,000 = $110,747.22

Debt ratio qualification

Formulas: $$Debt\ ratio = \frac{Housing\ expense + Other\ debt\ payment}{Monthly\ gross\ income}$$

Housing expense = (Monthly gross income x Debt ratio) - Other debt payments

Example: A lender uses a 36% debt ratio. A borrower earns $30,000 / year and has monthly non-housing debt payments of $500. What housing payment can she afford?

Housing expense = ($30,000 ÷ 12 x .36) - 500 = ($900 - 500) = $400

Investment

Appreciation Calculations

Simple appreciation

Formulas: *Total appreciation = Current value - Original price*

$$Total\ appreciation\ rate = \frac{Total\ appreciation}{Original\ price}$$

$$Average\ annual\ appreciation\ rate = \frac{Total\ appreciation\ rate}{Number\ of\ years}$$

$$One\ year\ appreciation\ rate = \frac{Annual\ appreciation\ amount}{Value\ at\ beginning\ of\ year}$$

Examples: 1.   A home purchased for $200,000 five years ago is now worth $300,000. What are the total appreciation amount, total appreciation rate, and average appreciation rate?

Total appreciation = ($300,000 - 200,000), or $100,000

Total appreciation rate = ($100,000 ÷ 200,000), or 50%

Average annual appreciation rate = 50% ÷ 5 years = 10%

2.   A home costing $250,000 is worth $268,000 one year later. What is the one-year appreciation rate?

One-year appreciation rate = ($18,000 ÷ 250,000) = 7.2%

Compounded appreciation

Formula: *Appreciated value = Beginning value x (1+ annual rate) x (1+ annual rate) ...... for the number of years in question*

Example: A $100,000 property is expected to appreciate 5% each year for the next 3 years. What will be its appreciated value at the end of this period?

Appreciated value = $100,000 x 1.05 x 1.05 x 1.05 = $115,762.50

Rate of Return, Investment Value, Income

Formulas: Where Income = net operating income (NOI); Rate = rate of return, cap rate, or percent yield; and Value = value, price or investment amount:

$$Rate = \frac{Income}{Value}$$

$$Value = \frac{Income}{Rate}$$

$$Income = Value \times Rate$$

Examples: 1.   An office building has $200,000 net income and sold for $3,200,000. What was the rate of return?

Rate = ($200,000 NOI ÷ 3,200,000 price) = 6.25%

2.   An office building has $200,000 net income and a cap rate of 6.25%. What is its value?

Value = ($200,000 ÷ 6.25%) = $3,200,000

3.   An office building sells for $3,200,000 at a cap rate of 6.25%. What is its NOI?

Income = $3,200,000 x 6.25% = $200,000

Basis, Adjusted Basis, and Capital Gain

Formulas: *Capital gain = Amount realized - Adjusted basis, where*

Amount realized = Sale price - Selling costs

Adjusted basis = Beginning basis + Capital improvements - Total depreciation

$$Total\ depreciation = \frac{Beginning\ depreciable\ basis}{Depreciation\ term\ (years)} \times Years\ depreciated$$

Depreciable basis = Initial property value + Capital improvements - Land value

Example: Tip: work example backwards from last formula to first formula.

An apartment building was purchased for $500,000, with the land value estimated to be $100,000. The owner added a $100,000 parking lot. The property was depreciated on a 40-year schedule (for present purposes!). Three years later the property sold for $700,000, and selling costs were $50,000. What was the capital gain?

1.   depreciable basis = $500,000 purchase price + 100,000 parking lot - 100,000 land = $500,000

2.   total depreciation = ($500,000 ÷ 40 years) x 3 years = $37,500

3.   adjusted basis = $500,000 purchase price + 100,000 parking lot - 37,500 total depreciation = $562,500

4.   amount realized = $700,000 sale price - 50,000 selling costs = $650,000

5.   capital gain = $650,000 amount realized - 562,500 adjusted basis = $87,500

Depreciation

Formulas: $$Annual\ depreciation = \frac{Beginning\ depreciable\ basis}{Depreciation\ term\ (number\ of\ years)}$$

Depreciable basis = (Initial property value + Any capital improvements - Land value)

Example: Property value = $500,000; land value = $110,000; depreciation term = 39 years

1.   ($500,000 - 110,000) = $390,000 depreciable basis

2.   ($390,000 ÷ 39 years) = $10,000 annual depreciation

Equity

Formula: *Equity = Current market value - Current loan balance(s)*

Example: A home that was purchased for $150,000 with a $100,000 loan is now worth $300,000. The current loan balance is $80,000. What is the homeowner's equity?

Equity = $300,000 value - $80,000 debt = $220,000

Net Income

Formula: *NOI = Potential rent - Vacancy loss + Other income - Operating expenses*

Note: NOI does not include debt payments!

Example: A building has 10 office suites generating annual potential rent of $10,000 each. Vacancy = 10% and annual expenses are $35,000. Vending machines yield $5,000. What is the NOI?

$100,000 rent - 10,000 vacancy + 5,000 other income - 35,000 expenses = $60,000 NOI

Cash Flow

Formula: *Cash flow = (Net Operating Income - Debt service) where debt service is PI payment*

Example: A building generates $100,000 NOI after expenses and has a debt payment of $40,000. What is its cash flow?

Cash flow = $100,000 - 40,000 = $60,000

Investment Property Income Tax Liability

Formula: *Tax liability = (NOI + Reserves - Interest expense - Depreciation) x Tax bracket*

Example: An office building has NOI of $200,000, an annual reserve expense of $20,000, interest expense of $130,000 and annual depreciation of $50,000. Assuming a 28% tax bracket, what is its income tax liability?

Tax liability = ($200,000 + 20,000 - 130,000 - 50,000) x 28% = $11,200

Return on Investment

Formula: $$ROI = \frac{NOI}{Price}$$

Example: An investment property generates a cash flow of $100,000 and appraises for $1,500,000. What is the owner's return on investment?

ROI = $100,000 ÷ 1,500,000 = 6.67%

Return on Equity

Formula: $$ROE = \frac{Cash\,flow}{Equity}$$

Example: An investment property generates a cash flow of $100,000. The owner has $500,000 equity in the property. What is the owner's return on equity?

ROE= $100,000 ÷ 500,000 = 20%

Real Estate Taxation

Converting Mill Rates

Definition: 1 mill = $.001; a mill rate of 1 mill per $1,000 = .1%; a 1% tax rate = 10 mills

Formula: $$Tax = \frac{Taxable\ value}{1,000} \times Mill\ rate$$

Example: A tax rate on a house with a $200,000 taxable value is 7 mills per thousand dollars of assessed valuation. What is the tax?

Tax = ($200,000 ÷ 1,000) x 7 mills = $1,400

Tax Base

Formula: *Tax base = Assessed valuations - Exemptions*

Example: A town has a total assessed valuation of $20,000,000 and exemptions of $4,000,000. What is the tax base?

$20,000,000 - 4,000,000 = $16,000,000

Tax Rate, Base, and Requirement

Formulas: $$Tax\ rate = \frac{Tax\ requirement}{Tax\ base}$$

$$Tax\ base = \frac{Tax\ requirement}{Tax\ rate}$$

Tax requirement = Tax base x Rate

Example: A town has a tax base of $160,000,000 and a budget of $8,000,000. What is the tax rate?

Tax rate = ($8,000,000 ÷ 160,000,000) = .05, or 5%, or 50 mills

29

Special Assessments

Formula: *Special assessment = Total special assessment cost x Homeowner's share*

Example: A homeowner owns 100' of an 800' seawall that must be repaired. The total assessment will be $80,000. What is the homeowner's assessment?

1.   Homeowner's share = 100' ÷ 800' = .125, or 12.5%

2.   Special assessment = $80,000 x 12.5% = $10,000

Commissions

Commission Splits

Formulas: *Total commission = Sale price x Commission rate*

Co-brokerage split = Total commission x Co-brokerage percent
Agent split = Co-brokerage split x Agent percent)

Broker split = Co-brokerage split - Agent split

Example: A $300,000 property sells at a 7% commission with a 50-50 co-brokerage split and a 60% agent split with her broker. What are total, co-brokerage, agent's, and broker's commissions?

Total commission = $300,000 x .07 = $21,000

Co-brokerage splits = $300,000 x .07 x .50 = $10,500

Agent split = $10,500 x .60 = $6,300

Agent's broker's split = $10.500 - 6,300 = $4,200

Seller's Net

Formula: *Seller's net = Sale Price - (sale price x commission) - Other closing costs - Loan balance*

Example: A home sells for $260,000 and has a loan balance of $200,000 at closing. The commission is 7% and other closing costs are $2,000. What is the seller's net?

Seller's net = ($260,000 - (260,000 x .07) - 2,000 - 200,000) = $39,800

Price to Net an Amount

Formula: *Sale Price = (Desired net + Closing costs + Loan payoff)) ÷ (1 - Commission rate)*

Example: A homeseller wants to net $50,000. The commission is 7%, the loan payoff is $150,000, and closing costs are $4,000. What must the price be?

Sale price = ($50,000 + 4,000 + 150,000) ÷ .93 = $219,355

Closing Costs, Prorations

30-Day 12-Month Method

Formulas:

$$\text{Monthly amount} = \frac{\text{Annual amount}}{12}$$

$$\text{Daily amount} = \frac{\text{Monthly amount}}{30}$$

Proration = (Monthly amount multiplied by the # months) + (Daily amount multiplied by the # days)

Example: An annual tax bill is $1,800. Closing is on April 10. What is the seller's share of the taxes?

    1.   Monthly amount = ($1,800 ÷ 12) = $150; no. of mounts = 3

    2.   Daily amount = ($150 ÷ 30) = $5.00; no. of days = 10

    3.   Proration = ($150 x 3) + ($5 x 10) = ($450 + 50) = $500 seller's share

---

365-Day Method

Formula: $$\text{Daily amount} = \frac{\text{Annual amount}}{365} \quad or \quad \frac{\text{Monthly amount}}{\text{Length of month}}$$

Proration = Daily amount multiplied by the # days

Example: An annual tax bill is $1,800. Closing is on April 10. What is the seller's share of the taxes?

    1.   Daily amount = ($1,800 ÷ 365) = $4.93

    2.   Jan 1 thru April 10 = (31 + 28 + 31 + 10) days, or 100 days

    3.   Proration = $4.93 x 100 days = $493 seller's share

Income Received in Advance (Rent)

Logic: *Credit buyer and debit seller for buyer's share*

Example: Seller receives $1,000 rent. The month is ¾ over.

    1.   Buyer's share is ($1,000 x 25%) = $250

    2.   Credit buyer / debit seller $250.

Expenses paid in Arrears (Tax)

Logic: *Credit buyer and debit seller for seller's share*

Example: Buyer will pay $1,000 taxes. The year is ¾ over.

1.   Buyer's share is ($1,000 x 25%) = $250

2.   Credit buyer / debit seller $750.

Insurance Coverage

Recovery with Co-Insurance Clauses

Formula:

$$Recovery = Damage\ claim \times \frac{\%\ Replacement\ cost\ covered}{Minimum\ coverage\ requirement}$$

Example: An owner insures a home for $100,000. Replacement cost is $150,000. A co-insurance clause requires coverage of 80% of replacement cost to avoid penalty. Fire destroys the house. What can the owner recover from the insurer?

Claim recovery = $150,000 x (67% cost covered ÷ 80% required) = $125,625

Math Questions

1. A lot measuring 20 and 2/3 acres must be divided equally between four heirs. How much will each heir receive?

   a. 5 1/3 acres
   b. 5 1/6 acres
   c. 2 2/3 acres
   d. 5 acres

2. A Realtor sells 7/8 of an acre for $40,000, and receives a 7% commission. If she splits with her broker 50-50, what did she receive per square foot?

   a. $.037 / SF
   b. $.074 / SF
   c. $.028/ SF
   d. $.058 / SF

3. A home appreciated 5 ¼% one year, then 6 2/3% the next year, then 7.2% the third year. What was the average appreciation over the 3-year period expressed as a decimal?

   a. 6.37%
   b. 6 1/3%
   c. 19.12%
   d. 6 2/3 %

4. A survey finds that agents in a particular region, on the average, receive 60% of ½ of 7% commissions on an average home sale price of $234,000. What is an agent's average commission in dollars?

   a. $16,380
   b. $3,276
   c. $9,828
   d. $4,914

5. Lots in the St. Cloud subdivision are selling for approximately $.50 / SF. The Andersons want to build a 2,500 SF home on a 1.5 acre corner lot. The custom builder can build the home for $135 / SF. What will the completed property cost the Andersons?

   a. $369,170
   b. $370,170
   c. $32,670
   d. $371,070

6.   Andre owned a ¼ acre lot. He wanted to construct a 120' x 80' tennis court on
       the lot. What approximate percentage of the lot will be left over, if any, when he
       has completed the construction?

       a. 12%
       b. 88%
       c. 3%
       d. 15%

7.   A developer wants to develop a 20-acre subdivision. He figures that the streets
       and common area will take up about 25% of this overall area. If the minimum lot
       size is to be 10,000 SF, how many lots can the developer have on this property?

       a. 64
       b. 653
       c. 65
       d. 87

8.   Jeannie leased an office building to a tenant under a 3-year lease for $20 / SF.
       Her commission is 4% of the rent payable over the lease term. If she received a
       $15,000 commission, how big was the office building?

       a. 18,750 SF
       b. 2,500 SF
       c. 6,250 SF
       d. 5,000 SF

9.   A homeowner wants to insulate the new recreation room in her basement. She
       has been told that 4" of insulation would do the job. The walls are all 8' high and
       respectively measure 15', 15', 17', and 17' in length. How many rolls will she
       need if each roll measures 4" x 2' x 50'?

       a. 6
       b. 50
       c. 5
       d. 3

10.   Kinza plans to mulch the flower area around her house. The house measures
       40' x 30', and she figures she'll mulch an area 8' in width to form a big rectangle
       all around the perimeter. She also figures 4" of depth should be sufficient. If a
       landscaper quotes Kinza that mulch costs $20 / cubic yard, how much will she
       spend?

       a. $918
       b. $1,019
       c. $260
       d. $340

11. Jamie Spiffup advised her clients they needed to paint their living room before showing the property. The walls of these rooms were all 8' high. The wall lengths were 14', 18', 16', and 18'. The job required two coats of paint. If a gallon of paint covers 200 SF, how many whole gallons would the homesellers have to buy?

a. 3
b. 4
c. 5
d. 6

12. A rancher has a rectangular 350 acre piece of property fronting a highway. He wants to put a barbed-wire fence along his road frontage. How much fence does he need if his property depth is 3,000'?

a. 3,904'
b. 5,082'
c. 508'
d. 1,852'

13. An investor just purchased a rectangular 2-acre retail lot for $250 a frontage foot. If she paid $100,000 total, what was the depth of the lot?

a. 400'
b. 250
c. 871'
d. 218'

14. A boat dock runs 80% of the length of a property's seawall. If the property is a rectangular ¼ acre lot with 80' of depth between the street and the seawall, how long is the dock?

a. 136'
b. 109
c. 64'
d. 124'

15. A homeowner paid $185,000 for a house three years ago. The house sells today for $239,000. How much has the property appreciated?

a. 23 %
b. 77 %
c. 29 %
d. 123 %

16. A property sells for $180,000 one year after it was purchased. If the annual appreciation rate is 10%, how much did the original buyer pay for it?

    a. $162,000
    b. $163,636
    c. $180,000
    d. $198,000

17. Yvonne bought a home in Biltmore Estates 3 years ago for $250,000. She obtained an 80% loan. Over the past three years, the total appreciation rate has been 18%, and she has paid down her loan by $4,000. What is Yvonne's equity after this period?

    a. $45,000
    b. $99,000
    c. $95,000
    d. $54,000

18. The roof of a property cost $10,000. The economic life of the roof is 20 years. Assuming the straight-line method of depreciation, what is the depreciated value of the roof after 3 years?

    a. $10,000
    b. $8,500
    c. $7,000
    d. $1,500

19. A property is purchased for $200,000. Improvements account for 75% of the value. Given a 39-year depreciation term, what is the annual depreciation expense?

    a. $3,846
    b. $5,128
    c. $6,410
    d. $8,294

20. Tim Bright had to report his home office depreciation for the tax year. He has a 2500 SF home and a 500 SF office area. Tim paid $280,000 for his home, and he figures the land portion carries about 25% of that value. If Tim depreciates on a 39-year basis, how much can he write off for his home office depreciation per year?

    a. $1,077
    b. $1,436
    c. $5,384
    d. $2,108

21. An investor paid $80,000 for a lot and $600,000 to have an apartment building constructed on it. He has depreciated the property for the past 10 years on a 39-year straight-line schedule. If he sells the property this year and realizes $780,000, what is his capital gain?

a. $253,846
b. $274,000
c. $100,000
d. $179,000

22. A homeowner bought a house five years ago for $250,000. Since then, the homeowner has spent $3,000 to pave the driveway and has added a central air-conditioning system at a cost of $4,000. What is the homeowner's adjusted basis if the house is sold today?

a. $256,000
b. $257,000
c. $244,000
d. $245,000

23. A homeowner sold her house and had net proceeds of $191,000. Her adjusted basis in the home was $176,000. She immediately bought another house for $200,000. What was her capital gain?

a. $191,000
b. $9,000
c. $15,000
d. None

24. A principal residence is bought for $180,000. A new porch is added, costing $7,000. Five years later the home sells for $220,000, and the closing costs $18,000. What is the homeowner's capital gain?

a. $15,000
b. $29,000
c. $33,000
d. $51,000

25. Mary Bright bought a home for $120,000, paying $24,000 down and taking a mortgage loan of $96,000. The following year she had a new roof put on, at a cost of $5,000. What is Mary's adjusted basis in the house if she now sells the house for $150,000?

a. $29,000
b. $96,000
c. $101,000
d. $125,000

26. A school district's tax rate is 10 mills. The school district's required revenue from taxes is $10,000,000. What is the tax base of the area?

   a. $10,000,000
   b. $100,000,000
   c. $1,000,000,000
   d. $100,000,000,000

27. A homeowner receives a tax bill that includes an amount for the library district, taxed at $1.00 per $1,000, and the fire protection district, taxed at $2.00 per $1,000. How much does the taxpayer have to pay for these two items if the property's taxable value is $47,000?

   a. $1,567
   b. $157
   c. $1,410
   d. $141

28. A town is replacing a sidewalk that serves five homes. The length of the sidewalk is 200 feet. Mary's property has 38 feet of front footage. If the cost of the project to be paid by a special assessment is $7,000, what will Mary's assessment be?

   a. $1,400
   b. $1,330
   c. $184
   d. $1,840

29. A homeowner's residence has an assessed valuation of $150,000, and a market value of $170,000. The homestead exemption is $25,000. Tax rates for the property are 7 mills for schools; 3 mills for the city; 2 mills for the county; and 1 mill for the local community college. What is the homeowner's tax bill?

   a. $1,625
   b. $1,885
   c. $1,950
   d. $2,210

30. The village of Parrish has an annual budget requirement of $20,000,000 to be funded by property taxes. Assessed valuations are $400,000,000, and exemptions total $25,000,000. What must the tax rate be to finance the budget?

   a. 4.70%
   b. 5.33%
   c. 5.00%
   d. 11.25%

31. Jacquie obtains a 75% LTV loan on her new $200,000 home with an annual interest rate of 6%. What is the first month's interest payment?

    a. $900
    b. $250
    c. $1,000
    d. $750

32. Ginnie has an interest-only home equity loan at an annual interest rate of 6%. If her monthly payment is $375, how much is the loan's principal balance?

    a. $22,500
    b. $75,000
    c. $6,250
    d. $46,500

33. The loan officer at FirstOne Bank tells Amanda she can afford a monthly payment of $1,300 on her new home loan. Assuming this is an interest-only loan, and the principal balance is $234,000, what interest rate is Amanda getting?

    a. 6.67%
    b. 5%
    c. 8.25%
    d. 6%

34. A borrower obtains a 30-year, amortized mortgage loan of $200,000 at 8%. What is the balloon payment at the end of the loan term if her PI payment is $1,466?

    a. 20,800
    b. $1,733
    c. $1,466
    d. Zero

35. A $300,000 loan has monthly interest-only payments of $2,000. Its annual interest rate is:

    a. 4%
    b. 6%
    c. 8%
    d. 10%

36. The Kruteks obtain a fixed-rate amortized 30-year loan for $280,000 @ 6.25% interest. If the monthly payments are $1,724, how much interest do the Kruteks pay in the second month of the loan?

    a. $1,748.33
    b. $1,456.95
    c. $1,458.33
    d. $1724.00

37. A lender offers the Amerines two alternative loan packages for their $60,000 home equity application. One option is an interest-only loan for 5 years @ 6.5% interest with no points, and the second, a 6.25% interest-only loan for 5 years with 1 point to be paid at closing. Which loan will cost the Amerines less total interest, and by how much?

    a. The first option, by $150.
    b. The second option, by 150.
    c. The second option, by $750.
    d. Both options charge the same amount of interest.

38. Carrie Creative pays $650 per month for her mortgage. One day she sells her home under a contract for deed. Her bargain price is set at 90% of the appraised value, and Carrie offers to give the buyers an 80% interest-only loan for 7% with a 4-year balloon. If the appraisal came in at $300,000, and the bank allows her to keep her underlying loan, what is Carrie's monthly net profit?

    a. $1,260
    b. $610
    c. $650
    d. $750

39. Sally paid $200,000 for her house 3 years ago. She believes it has appreciated 5% each year, and now wants to sell. At the end of the third year, the home appraises out at 90% of Sally's projection of current market value, but Sally insists on listing the property at the price indicated by the appreciation rate. The buyer's lender agrees to a 90% loan-to-appraised-value loan. Given these circumstances, how much down payment will the buyer need to make to buy the property?

    a. $12,465
    b. $43,990
    c. $41,895
    d. $23,152

40. Greg recently obtained an 80% loan on his $320,000 home, and he had to pay $4,480 for points. How many points did he pay?

a. 1.4 points
b. 1.75 points
c. 4.48 points
d. 3.584 points

41. Craig is buying Warren's house for $300,000. Craig's loan amount is $225,000. He has agreed to pay 1.5 points at closing. How much will Craig pay for points?

a. $375
b. $4,500
c. $3,375
d. $337.50

42. A lender charges 2 points on a 5-year, $100,000 loan carrying a 6% interest rate. As an alternative, she will make the same loan with no points, but wants to break even. What interest rate will she have to charge?

a. 8%
b. 3.2%
c. 6.4%
d. 7%

43. A homebuyer is low on cash but has a strong monthly income. The lender, who normally charges 1.5 points for a 5.75% loan, agrees to add the points charge to the loan balance. If the loan is interest-only for $200,000, how much will this increase the monthly payment?

a. $150
b. $17
c. $15
d. $144

44. A lender determines that a homebuyer can afford to borrow $120,000 on a mortgage loan. The lender requires an 80% loan-to-value ratio. How much can the borrower pay for a property and still qualify for this loan amount?

a. $96,000
b. $106,000
c. $150,000
d. $160,000

45. Annika can afford to spend $4,000 in closing costs to refinance her home. The lender quotes closing costs of $800 plus 2 points. The house appraised out at $240,000, and she can get an 80% loan. Can Annika afford to refinance?

   a. No, she is short by $64.
   b. No, she is short by $1,600.
   c. No, she is short by $640.
   d. Yes, she in fact breaks even.

46. A lender offers an investor a maximum 70% LTV loan on the appraised value of a property. If the investor pays $230,000 for the property, and this is 15% more than the appraised value, how much will the investor have to pay as a down payment?

   a. $93,150
   b. $79,350
   c. $90,000
   d. $69,000

47. A home buyer pays $1,600 / month for the interest-only loan on his new house. The loan's interest rate is 6.75%. If she obtained a 75% loan, what was the purchase price?

   a. $213,333
   b. $31,604
   c. $379,259
   d. $256,000

48. Misty has monthly loan payments of $1,264. Her loan is for $200,000 @ 6 ½% interest. How much principal does she pay in the first month of her payment schedule?

   a. $1,083
   b. $1,264
   c. $181
   d. $191

49. A loan applicant has an annual gross income of $36,000. How much will a lender allow the applicant to pay for monthly housing expense to qualify for a loan if the lender uses an income ratio of 28%?

   a. $2,160
   b. $840
   c. $1,008
   d. $720

50. Assume FHA qualifies borrowers based on a 41% debt ratio, meaning that the borrower's total monthly debt including the loan, taxes, insurance and non-housing debt cannot exceed 41% of the borrower's monthly income. If a borrower grosses $4,000 per month and pays $600 monthly for non-housing debt obligations, what monthly payment for housing expenses can this person afford based on this ratio?

a. $760
b. $1,394
c. $1,040
d. $1,404

51. A borrower earns $3,000/month and makes credit card and car note payments of $500. A conventional lender requires a 27% income ratio. What monthly amount for housing expenses (principal, interest, taxes, insurance) will the lender allow this person to have in order to qualify for a conventional mortgage loan?

a. $810
b. $675
c. $972
d. $1,040

52. Thomas wants to buy a $270,000 home. He makes $60,000 / year and has $40,000 to put down. The lender uses a 28% income ratio to qualify borrowers. Assuming an interest-only loan @ 6.5%, does Thomas qualify to buy the house, and, secondly, by what monthly surplus or shortfall?

a. He can afford it by a $154 margin.
b. He can afford it by a $1,850 margin.
c. He cannot afford it by a $154 margin.
d. He cannot afford it by a $1850 margin.

53. A property is being appraised using the income capitalization approach. Annually, it has potential gross income of $30,000, vacancy and credit losses of $1,500, and operating expenses of $10,000. Using a capitalization rate of 9%, what is the indicated value (to the nearest $1,000)?

a. $206,000
b. $167,000
c. $222,000
d. $180,000

54. If gross income on a property is 30,000, net income is $20,000 and the cap rate is 5%, the value of the property using the income capitalization method is

a. $600,000
b. $400,000
c. $6,000,000
d. $4,000,000

55. A certain investor wants an 11% return on investment from any real estate investment. A property priced at $360,000 has gross income of $60,000 and expenses of $22,000. Approximately how much too high or too low is the price of this property for the investor to obtain her desired return exactly?

a. $1,000 overpriced.
b. $8,000 underpriced.
c. $15,000 overpriced.
d. $16,000 underpriced.

56. A commercial property sold recently for $500,000. The property had an NOI, or net income, of $25,000. What was the capitalization rate at which this property sold?

a. 20%
b. 2%
c. 5%
d. 15%

57. An office building investor sees a listing of an office building which is priced at $2 million. He loves the property, but he knows he needs to make a return of at least 8% to satisfy his partners. If the building is 25,000 SF, rents for $10/SF per year, has 5% vacancy, and annual expenses of $70,000, should he buy it? What is his return?

a. No, since he will yield 2.00%.
b. Yes, since he will yield 8.375%.
c. Yes, since he will yield 8%.
d. Yes, since he will yield 9%.

58. Johnnie Holiday wants to sell his 10-unit resort hotel and become a millionaire. The rooms rent for $150 per night and he figures there are 200 rentable nights per year. His expenses are 33% of his gross income. Johnnie wants a price of $1 million, but he knows vacancy is the big challenge. He finds a buyer who will buy the resort if it returns 10%. What must Johnnie's maximum vacancy be to satisfy his buyer?

a. 25%
b. 40%
c. 50%
d. 60%

59. A property is being appraised by the cost approach. The appraiser estimates that the land is worth $10,000 and the replacement cost of the improvements is $75,000. Total depreciation from all causes is $7,000. What is the indicated value of the property?

a. $68,000
b. $92,000
c. $82,000
d. $78,000

60. An apartment owner paid $500,000 for her complex 5 years ago. An appraiser at that time valued the land @ $100,000, but land has appreciated 25% over this period. The investor has used a 40-year straight-line depreciation method to depreciate the property. What is its current value using the cost approach?

a. $437,500
b. $462,500
c. $475,000
d. $546,875

61. An apartment building that sold for $450,000 had monthly gross rent receipts of $3,000. What is its monthly gross rent multiplier?

a. 12.5
b. .01
c. .08
d. 150

62. A rental house has monthly gross income of $1,200. A suitable gross income multiplier derived from market data is 14.1. What estimated sale price (to the nearest $1,000) is indicated?

a. $169,000
b. $102,000
c. $203,000
d. $173,000

63. A 20-unit apartment building grosses $15,000 / month. If the market value of this property is $2,200,000, what is its gross income multiplier?

a. 14.66
b. 122
c. 12.22
d. .08

64. A property has sold for $127,000. The listing agreement calls for a commission of 7%. The listing broker and selling broker agree to share the commission equally. What will the listing agent receive if the agent is scheduled to get a 40% share from his broker?

a. $4,445
b. $3,556
c. $2,667
d. $1,778

65. A real estate salesperson brings a buyer to a For-Sale-By-Owner transaction. The home sells for $245,000, and the seller agrees to pay a commission of 3%. The salesperson is on a 65% commission schedule with her broker, who pays her the 65% minus office expenses of $500. How much will the salesperson receive from this transaction?

a. $4,778
b. $4,452
c. $4,278
d. $3,175

66. Agent Stephanie sells Monica's house, her own listing, for $287,500. The brokerage commission is 6%. The co-brokerage split is 50-50, and Stephanie gets 60% of her broker's commission. How much will Stephanie's broker receive?

a. $8,625
b. $6,900
c. $3,450
d. $10,350

67. Spenser, who works for selling broker Smith, sells a house listed by listing broker Adams. The house sells for $425,000. The co-brokerage split between Smith and Adams is 50-50. Spenser is on a 65% commission schedule with Smith. If the total commission rate is 7%, what is Spenser's commission?

a. $9,669
b. $13,812
c. $14,875
d. $19,338

68. Ryan sells his home for $210,000. He must pay a 7% brokerage fee which will be split evenly between the selling broker and listing broker. He will also incur $3,000 in other closing costs and pay off a $165,000 mortgage balance. How much can Ryan expect to receive at closing?

a. $27,300
b. $34,650
c. $32,300
d. $23,700

69. Agent Howard sold his client's farm acreage for $620,000 in a co-brokered deal. The seller's non-commission closing costs were $6,000, and she netted $570,000 at closing. If Howard split evenly with his broker, and the co-brokerage was 50-50, what was the commission rate?

   a. 7%
   b. 3.5%
   c. 5.3%
   d. Cannot be determined from the data provided.

70. Agent Graham sold his client's home for $320,000 through a cooperating buyer agent. The seller's non-commission closing costs were $3,000. At the closing, Graham received $6,240 and the seller netted $296,200. If Graham gets a 60% split of his broker's commission, and the co-brokerage was 50-50, what was the commission rate?

   a. 6.5%
   b. 3.25%
   c. 7%
   d. 3.9%

71. Hannah's goal is to make $50,000 in commissions next year. Her broker agrees to split 50-50 with her on her first $1,000,000 in sales, and raise her split to 60-40 thereafter. If the average commission charged sellers is 7%, and Hannah assumes that all deals will be co-brokered 50-50, how much real estate must she sell to reach her goal?

   a. $2,547,619
   b. $2,321,428
   c. $1,547,619
   d. $5,095,238

72. William sold $4 million in real estate last year and expects to do 25% better this year. His broker offers William two alternative commission packages. She will either give him Plan A, a 65-35 split with no desk fee, or Plan B, an 80-20 split with a $1,000/month desk fee. If commissions average 7%, and all the closings are co-brokered 50-50, which package would William take to make the most money? What is the margin between the two?

   a. Plan A is better by $15,240.
   b. Plan A is better by $6,700.
   c. Plan B is better by $7,125.
   d. Plan B is better by $14,250.

73. Darlene sold two homes as a buyer rep for $160,000 each @ 6% overall commission with a 50-50 co-brokerage split; then she sold 3 of her own listings for a total of $640,000 @ 7% overall commission; then sold her 50-acre land tract listing for $250,000 @ 10% overall commission. After her broker's split of 40%, what did Darlene make?

    a. $48,600
    b. $41,760
    c. $40,410
    d. $47,640

74. A seller paid a $500 annual insurance premium in advance. The closing takes place on the 90th day of the year and belongs to the seller. Using the 360-day method, what should appear on the closing statement?
    a. A debit to the buyer and credit to the seller for $375.
    b. A debit to the buyer for $375 and a credit to the seller for $125.
    c. A debit to the buyer for $125 and a credit to the seller for $375
    d. A debit to the buyer and credit to the seller for $125.

75. A seller received a monthly rental payment of $1,000 in advance. At closing, the seller has earned only $320 of this rent. What should appear on the closing statement?

    a. A debit to the seller and credit to the buyer for $320.
    b. A debit to the seller for $680 and a credit to the buyer for $320.
    c. A debit to the seller for $320 and a credit to the buyer for $1,000.
    d. A debit to the seller and credit to the buyer for $680.

76. A buyer will receive a utilities bill for an estimated $300 at the end of the month. At closing, the seller has used an estimated $100 of the bill. What should appear on the closing statement?

    a. A debit to the seller and credit to the buyer for $200.
    b. A debit to the seller and credit to the buyer for $100.
    c. A debit to the buyer and credit to the seller for $200.
    d. A debit to the buyer and credit to the seller for $100.

77. A sale transaction closes on April 1, the ninety-first day of the tax year. The day of closing belongs to the seller. Real estate taxes for the year, not yet billed, are expected to be $3,150. According to the 365-day method, what should appear on the closing statement?

    a. A debit to the buyer and credit to the seller for $2,364.62.
    b. A debit to the buyer and credit to the seller for $785.34.
    c. A credit to the buyer and debit to the seller for $785.34.
    d. A credit to the buyer and debit to the seller for $2,364.62.

49

78. A sale transaction closes on July 4. The day of closing belongs to the seller. On January 1, the seller paid a hazard insurance premium of $375 for the calendar year. According to the 12-month/30-day method, what should appear on the closing statement?

   a. A debit to the buyer and credit to the seller for $191.67.
   b. A debit to the buyer and credit to the seller for $183.33.
   c. A credit to the buyer and debit to the seller for $183.33.
   d. A credit to the buyer and debit to the seller for $191.67.

79. Adam is buying Kiki's house. The closing date (day belongs to seller) of the sale transaction is September 1 (day 244 of the year). Current Year real estate taxes are $1,100 (will be billed to buyer next year). Use the 365-day method for prorating. What is the seller's share of the real estate taxes?
   a. $364.66
   b. $367.67
   c. $732.33
   d. $735.34

80. Alexis is buying Morton's house. The closing date (day belongs to seller) of the sale transaction is September 1 (day 244 of the year). Her loan has a monthly payment of $577.84, with $525 going to interest in the first month. At closing, Alexis must pre-pay interest for the period of Sept. 2-Sept. 30. Use the 365-day method for prorating. What is her prepaid interest amount?

   a. $507.50
   b. $525.00
   c. $543.10
   d. $558.58

81. A sale transaction on rental property closes on December 16. The landlord received the December rent of $713 on December 1. Assuming the closing day is the buyer's, and that the 365-day method is used for prorating, which of the following entries would appear on the settlement statement?

   a. Debit seller $345.00.
   b. Credit seller $713.00.
   c. Debit buyer $345.00.
   d. Credit buyer $368.00.

82. A home sells for $322,600 in Central County. Here, transfer taxes are set at $1.00 per $500 of the sale price. Title insurance runs $450, and the attorney costs $550. The Realtor's commission is 7%, and the mortgage balance is $210,000. Annual real estate taxes are estimated to be $4,000, half of which will have to be charged to the seller. If the seller pays all of these expenses, what will she net at closing?

    a. $81,872
    b. $87,947
    c. $84,572
    d. $86,372

83. Home buyer Henry has contracted to buy a $292,000 home. He has put $5,000 down on the property, and plans to get an 80% loan. The lender will be charging 1.5 points. Other closing fees will be $1,200. Annual real estate taxes will be $3,600, 25% of which are the seller's responsibility. Additionally, Henry's homeowner's policy will run $650. How much should Henry bring to closing? Hint: Property taxes are paid in arrears by the buyer.

    a. $61,054
    b. $62,854
    c. $56,054
    d. $57,854

84. A farmer wants to net at least $5,000/acre on the sale of his 300-acre property. If he allows for 10% commissions and closing costs, and to allow for negotiating room, he wants to get 95% of the listing price as the selling price, what should his listing price be per acre?

    a. $5,750
    b. $5,882
    c. $4,250
    d. $5,848

85. Courtney wants to net $70,000 on her house. Her closing costs will be 3,000 plus a 6% commission. She owes $150,000 on her loan. What should the sale price be?

    a. $236,380
    b. $237,234
    c. $236,200
    d. $235,000

86. Alexandra wants to make a 20% profit on her $40,000 land investment. She figures agents charge a 10% commission, and that closing costs will be an additional $1,500. What should she accept as a final sale price?

    a. $54,300
    b. $55,000
    c. $54,450
    d. $52,800

87. A house is being appraised using the sales comparison approach. The house has three bedrooms, two bathrooms, and a patio. The appraiser selects a comparable house that has three bedrooms, 2.5 bathrooms, and no patio. The comparable house just sold for $100,000. A half-bath is valued at $5,000, and a patio at $1,000. Assuming all else is equal, what is the adjusted value of the comparable?

a. $100,000
b. $104,000
c. $96,000
d. $106,000

88. In a comparative market analysis, the subject property has 4 bedrooms, 2 baths, a ¾ acre lot and a swimming pool. A $320,000 comparable has 4 bedrooms, 3 baths, no pool, a ¼ acre lot, and a screened porch. The appraiser adjusts by $5,000 for the bath, $15,000 for the pool, $13,000 for the lot size difference and $3,500 for the porch. What is the indicated value of the subject?

a. $326,500
b. $300,500
c. $339,500
d. $320,000

89. Bill Holdfast owns a small retail property that he inherited from his father. There are no mortgages or interest expenses connected with the property. Bill takes an annual cost recovery expense of $5,000. The property has a monthly gross income of $1,500 and monthly operating expenses of $500. Bill's taxable income from this property will be taxed at a rate of 30%. What is the tax liability for the year?

a. $2,100
b. $3,600
c. $3,900
d. $7,000

90. A property has a net income of $50,000, interest payments of $35,000, principal payments of $3,000, and annual cost recovery of $7,000. The property's tax rate is 28%. What is the property's annual tax on income?

a. $4,200
b. $3,360
c. $2,240
d. $1,400

91. A family purchased a $90,000 lot to build a custom home. At the date of closing, the lot was assessed at $84,550 and the tax rate was $1.91 / $100 assessed valuation. When they completed the home, the assessment increased by $235,000 to include the new construction. If the monthly tax escrow is based on the assessed value, what will the monthly tax escrow be?

    a. $517
    b. $6096
    c. $508
    d. $367

92. An investor bought 3 oversized lots in order to subdivide. He paid $40,000 for the lots. After subdividing, the investor was able to sell each lot for $18,000. Excluding commissions and closing costs, what per cent profit did the investor realize?

    a. 145%
    b. 45%
    c. 68.9%
    d. 35%

93. Charlene bought a property with an area of 174,240 SF @ $1.25 per square foot. She then subdivided the parcel into ¼ acre lots, allowing for a 25% loss factor for roads. What price must each lot sell for to net Charlene 25% profit?

    a. 22,787
    b. $24,200
    c. $18,150
    d. $17,015

94. LeBron and James purchase a $320,000 home and consider a 90% loan for 30 years @ 6.5% interest. The monthly loan constant for this loan is 6.3207. An alternative 20 year loan @6 3/8 % can also be obtained. Its monthly loan constant is 7.3824. How much more interest will the 30-year loan cost over the life of the loan? (Round monthly payments to the nearest dollar.)

    a. $8,280
    b. 233,920
    c. $367,200
    d. $144,960

95. The Bankhead family purchased a home for $180,000 five years ago and obtained an 80% LTV loan. Now the property has appreciated 25%. In addition, the loan has been paid down $11,000. What is the Bankhead's current equity in the home?

    a. $47,000
    b. $81,000
    c. $45,000
    d. $92,000

96. A mortgage lender uses an income qualification ratio of 28% for the monthly PITI payment. The Poormons earn $82,000 per year. Assuming that interest rates are 7% on 30-year loans, the monthly loan constant is 6.6531. If taxes are estimated to be $6,000 and insurance $1,200, how much can the Poormons borrow?

    a. $287,535
    b. $197,532
    c. $292,857
    d. $197, 352

97. Marcos Pizza has a percentage lease on its 1,500 SF space in the Asheville Center. The terms are $1.25 / SF / month rent plus 2% of the store's gross income. If monthly sales averaged $35,000 last year, how much annual rent did Marcos Pizza pay last year?

    a. $30,900
    b. $2,575
    c. $10,275
    d. $8,400

98. A commercial builder has a downtown lot with 250 frontage feet on Broadway. The lot is 200' deep. By code, the builder must allow for a 15' setback on Broadway, and 10' on both sides of the lot. How many square feet does the builder have left to build on?

    a. 44,400 SF
    b. 42,550 SF
    c. 42,250 SF
    d. 50,000 SF

99. Retailer Bernie owns a commercial store front lot that he wants to sell. The lot is .75 acres with a depth of 250' of depth on either side. The agent informs Bernie he can get $1,250 per front foot for the lot, but to ask for 15% more, rounded to the nearest thousand, to allow room for negotiation. What total price should Bernie list the property for?

    a. $360,000
    b. $188,000
    c. $163,000
    d. $250,000

100. Aaron finally found a buyer for his six-plex and closing is set for June 20th. At closing, four of his tenants have paid their $650 rent and two of the units remain unoccupied. What will the proration be assuming the 365-day method and that the closing day belongs to the seller?

a. Debit seller, credit buyer $1,733.30.
b. Debit seller, credit buyer $866.70.
c. Credit seller, debit buyer $866.70.
d. Credit seller, debit buyer $1,733.30.

101. George and Mary have owned a rental house for 10 years. They bought it for $240,000 and estimated the land value @ 25%. If the property is depreciated on a 39-year schedule, and appreciation totals 50% over the period, what is their gain if they sell the property today?

a. $159,230
b. $166,150
c. $181,538
d. $120,000

102. Bonny agrees to pay off her home equity loan with a single payment of $8,900 in 1½ years. If her loan is for $8,000, what is her annual interest rate?

a. 7.5%
b. 11.25%
c. 6.74%
d. 13.3%

103. The Jacksons have purchased a $320,000 home. The land is worth 25%, and they insure the improvements @ 75% of their replacement value. If the Jacksons suffer damage estimated at $200,000, and they have an 80% co-insurance clause, what will their recovery be from the policy?

a. Zero
b. $200,000
c. $240,000
d. $187,500

104. The Harrisons carry a $140,000 property insurance policy which covers 75% of the replacement cost of their insurable property, valued at $190,000. They have an 80% co-insurance requirement in the policy. If the family incurs a $150,000 loss, what if any amount will the Harrisons recover?

a. $159,999
b. $140,625
c. $140,000
d. $187,500

105. Skip Billips is set to close on the sale of his home at the end of next month, on December 31. Skip has an overdue personal property tax bill of $250 which was payable September 1st. At closing, the entry for the tax bill will be

a. a debit of $250 to the seller.
b. a credit of $250 to the seller.
c. a debit of $250 to the seller and a credit of $250 to the buyer.
d. a debit to the seller and credit to the buyer for $166.67.

106. George is closing on his home purchase March 10. The purchase price is $360,000, and George is assuming the seller's loan of $200,000. What are the entries at closing for George's loan assumption.

a. Debit seller and credit buyer $200,000.
b. Credit seller and debit buyer $200,000.
c. Debit seller $200,000 only.
d. Credit buyer $200,000 only.

107. Ken and Tammy are closing on their home sale next month. The sale price was $260,000, and the loan payoff amount is $190,000. What is the closing statement entry for the loan payoff?

a. A credit of $70,000 to the sellers.
b. A debit of $70,000 to the sellers.
c. A debit of $190,000 to the sellers.
d. A credit of $190,000 to the sellers.

108. Sheryl's home is valued at $250,000. She has insurance coverage of $160,000 with an 80% co-insurance clause. If Sheryl has a damage claim amounting to $100,000, how much will she receive from her policy?

a. $32,000
b. $60,000
c. $80,000
d. $100,000

109. Karen obtains a new loan @ 75% of her home's price of $400,000. Her interest rate is 6% and the monthly loan constant is 6.443. How much principal does Karen pay in the first month's payment?

a. $132.90
b. $321.30
c. $400.00
d. $432.90

110. Brittany bought a house five years ago for $220,000 and obtained an 80% loan. Now the home is worth $360,000, and her loan balance has gone down $14,000. What is her average equity gain per year as a percent?

a. 70%
b. 450%
c. 31%
d. 350%

111. Homebuyer Betty applies for a $200,000 loan with a monthly payment of $1,289. The lender is underwriting with qualifying ratios of 28% for the income ratio, and 36% for the debt ratio. Betty's monthly income is $5,600. Her monthly car payments are $400, and her revolving credit payments are $525/month. Which of the following is true?

a. Betty qualifies under both ratios.
b. Betty qualifies under the income ratio, but not the debt ratio.
c. Betty qualifies under the debt ratio, but not the income ratio.
d. Betty does not qualify under either ratio.

112. Calculate how many acres are in the Southwestern ¼ of the Northern ½ of the Eastern ½ of Section 14.

a. 20 acres
b. 40 acres
c. 60 acres
d. 5 acres

113. Homeowner Theresa owns the Northwestern ¼ of the Northwestern ¼ of the Northwestern ¼ of Section 4. How many acres is that property?

a. 4 acres
b. 40 acres
c. 10 acres
d. 8 acres

114. Seller Tyronne receives an offer of $689,000 on a property he listed at $749,000. How much is the offer as a percent of the listing price?

a. 87%
b. 89%
c. 92%
d. 109%

115. Seller Andy requires a 3.5% deposit on all offers. Buyer Josh wants to offer $312,000 for the property. The property was appraised at $325,000. What must the earnest money deposit be if Josh presents his current offer?

a. $10,920
b. $11,375
c. $9,500
d. $10,538

Answer Key

1.   (b)   5 1/6 acres

20 2/3 acres = 62/3rds acres. 62/3rds / 4 is the same as 62/3 x ¼, or 62/12. Reducing this fraction, we get 31/6ths, or 5 1/6 acre. <percentages, decimals, fractions 1>

2.   (a) $.037 / SF

7/8ths of an acre = (7 x 43,560 SF) / 8, or 38,115 SF. Her commission was (.07 x $40,000) x .50, or $1,400. $1,400 / 38,115 SF = $.037 / SF. <percentages, decimals, fractions 2>

3.   (a)   6.37%

First, convert to decimals: 5 ¼ % = 5.25%; 6 2/3% = 6.67. Thus total appreciation = (5.25% + 6.67% + 7.2%), or 19.12%. Divide by 3 to derive the average: 19.12% / 3 = 6.37% <percentages, decimals, fractions 3>

4.   (d) $4,914

Take 7% of $234,000, or $16,380. Take out half of this, leaving $8,190. Then take 60% of this amount, or ($8,190 x .60) = $4,914. <percentages, decimals, fractions 4>

5.   (b) $370,170

The land costs $.50 x 43,560 SF/ac. x 1.5 ac, or $32,670. The home will cost $135 / SF x 2,500, or $337,500. The total property will cost $370,170. <area measurement 1>

6.   (a)   12%

The lot measures 43,560 / 4, or 10,890 SF. The tennis court will take up 9,600 SF, leaving 1,290 SF. This amount is 11.8% of the total lot area. <area measurement 2>

7.   (c)   65

The total area available for lots is 15 acres, or 653,400 SF (15x 43,560). Dividing this area by 10,000 SF / lot = 65.34. Thus he can have a 65-lot subdivision. <area measurement 3>

8. **(c)** **6,250 SF**

If Jeannie received $15,000 for three years, she averaged $5,000 for each year. Thus $5,000 = 4% times ($20 X) where X is the building area. $5,000 = .04 x 20 X. $5,000 /.8 = 6,250 SF <area measurement 4>

9. **(a)** **6**

First, the requirement = 2(15'x 8') +2 (17'x 8') =512 SF. Each roll is 2'x 50', or 100 SF. Thus she will need 6 rolls. <area measurements 5>

10. **(d)** **$340**

First figure the area to be mulched. If the home is 40 x 30, the flower area adds 8' to each side of the house. Thus the outside perimeter of the flowered area is (40+8+8) by (30+8+8), or 46' by 56'. The area of the flowered area is (46' x 56') minus the house area of 1,200 SF. Thus the flowered area is 1,376 SF. Second, figure the volume of the mulch: 1,376 SF x 4" = 459 cubic feet. Since there are 27 cubic feet in a cubic yard, she will need 459 / 27 cubic yards, or 17. Thus she will spend $340. <area measurements 6>

11. **(d)** **6**

The total area of the living room is (8' x 14' + 8' x 18' + 8' x 16' + 8' x 18') = 528 SF. Painting two coats doubles this to 1056 SF. They will therefore need 1,056 / 200 SF, or 6 whole gallons. <area measurements 7>

12. **(b)** **5,082'**

This property consists of (350ac x 43,560 SF / acre), or 15,246,000 SF. Dividing this by the depth of the parcel will yield the road frontage length: 15,246,000 / 3,000 = 5082' fencing required. <area measurement 8>

13. **(d)** **218'**

Since the investor paid $100,000 total, and that equals $250 per frontage foot, there are 400 frontage feet (100,000 / 250). If the property is two acres, it totals 87,120 SF. Dividing this by 400 produces a lot depth of 217.8'. <area measurements 9>

14. **(b)** **109**

First, the lot is ¼ acre, or 43,560 SF / 4, or 10,890 SF. If the lot is 80' deep, divide the depth into the lot area to calculate the length of the seawall: 10,890 / 80 = 136.12'. Multiply this length times 80% to derive the length of the dock: .80 x 136' = 109'. <area measurements 10>

15. (c) 29 %.

Appreciation as a per cent can be estimated by (1) subtracting the estimated current market value from the price originally paid (239,000 - 185,000 = 54,000) and (2) dividing the result by the original price (54,000 / 185,000 = . 29 or 29%). >appreciation 1<

16. (b) $163,636

The selling price is 110% of the purchase price. Therefore, the purchase price is the selling price divided by 1.1 (110%). $180,000 / 1.1 = $163,636. >appreciation 2<

17. (b) $99,000

Equity = Market value – loan balance. Her loan was .8 x $250,000, or $200,000. Her initial equity was 20% x $250,000, or $50,000. The home has appreciated 18%, or (.18 x $250,000), or 45,000. Plus she has paid down her loan $4,000. Her equity is therefore ($295,000 market value – 196,000 loan), or $99,000. <appreciation 3>

18. (b) $8,500.

First derive the annual depreciation which is the cost divided by the economic life. Then multiply annual depreciation times the number of years to identify total depreciation. Remember to subtract depreciation from the original cost if the question asks for the ending value. Thus, ($10,000 / 20 years x 3 years) = $1,500 total depreciation. The ending value is $10,000 – 1,500, or $8,500. >depreciation 1<

19. (a) $3,846

Since only the improvement portion of the property can be depreciated, the depreciable basis is $200,000 x 75%, or $150,000. The annual depreciation expense is $150,000 / 39 years, or $3,846. >depreciation 2<

20. (a) $1,077

First, Tim's depreciable basis, without the land, is $280,000 x 75%, or $210,000. The annual depreciation for the entire home is ($210,000 / 39 years), or $5,384.61. Second, his office is 20% of the house (500 sf / 2,500 sf). Therefore Tim can take annual depreciation of ($5,384.61 x 20%), or $1,076.92. <depreciation 3>

21. (a) $253,846

Total depreciation on this property = ($600,000 / 39 years) x 10 years, or $153,846. His adjusted basis is therefore ($680,000 original price – 153,846 depreciation taken), or $526,154. The gain is then ($780,000 – 526,154), or $253,846. <depreciation 4>

22.   (b) $257,000.

Adjusted basis = beginning basis ($250,000) + capital improvements ($3,000 + $4,000) – depreciation (0) = adjusted basis ($257,000). >cap gain and basis 1<

23.   (c) $15,000.

Capital gain = amount realized (net sales proceeds, $191,000) - adjusted basis ($176,000) = ($15,000). >cap gain and basis 2<

24.   (a) $15,000

Gain = amount realized ($220,000 - 18,000) - adjusted basis ($180,000 + $7,000) = ($15,000). >cap gain and basis 3<

25.   (d) $125,000)

The basic formula for adjusted basis is: Beginning Basis + Capital Improvements - Exclusions and Credits = Adjusted Basis. Mary's adjusted basis is therefore $120.000 + $5,000 = $125,000. The financing terms and subsequent selling price are not relevant. >cap gain and basis 4<

26.   (c) $1,000,000,000.

The mill rate = (tax requirement / the tax base). A mill is one one-thousandth of a dollar ($.001). To solve for the tax base, reconfigure this formula to be: Base = Tax Requirement / Mill Rate. Thus the Base = $10,000,000 / .010, or $1,000,000,000. >taxation 1<

27.   (d) $141.

The library tax is ($1.00 x 47) and the fire district tax is ($2.00 x 47). Thus the tax is $141. >taxation 2<

28.   (b) $1,330.

Special assessments are based on the cost of the improvement and apportioned on a pro rata basis among benefiting properties according to the value that each parcel will receive from the improvement. Here, Mary's share is 38 / 200, or 19%. 19% x $7,000 = $1,330.   >Taxation 3 <

29.   (a) $1,625

First, always use the assessed valuation, not the market value. Subtract out the homestead exemption to derive taxable value, or $150,000 − 25,000 = $125,000. As a shortcut to calculating the tax bill, simply add up all the mills, multiply them times .001 to convert mills to decimals, then multiply this number times the taxable value. Thus (7 + 3 + 2 + 1) x .001 x $125,000 = $1,625. >taxation 4<

30.   (b)   5.33%

The rate = budget / tax base. Thus, $20,000,000 / (400,000,000 − 25,000,000) = 5.33% >taxation 5<

31.   (d) $750

The loan amount is $200,000 x .75, or $150,000. The first month interest equals ($150,000 x 6%) / 12 months, or $750. >Rates, payments 1<

32.   (b) $75,000.

The equation for the loan amount is (annual interest divided by the interest rate) = loan amount. Thus, ($375 x 12) / .06 = $75,000. >rates, payments 2<

33.   (a)   6.67%.

The equation for the interest rate is (annual payment / loan amount) = interest rate. Thus ($1,300 x 12) / $234,000 = 6.67%.   >rates, payments 3<

34.   (d)   Zero.

If a loan is fully amortizing, its loan balance is zero at the end of the loan term. >rates, payments 4<

35.   (c)   8%.

Remember that (Principal x Rate) = Annual Payment. If you know 2 of the 3 quantities, you can solve for the third. To solve for the interest rate, change the formula to (Rate = Annual Payment / Principal). Thus ($2,000 x 12 months) / $300,000 = .08, or 8%. >rates, payments 5<

36.   (b) $1,456.95

In the first month they pay interest of ($280,000 x 6.25%) / 12, or $1,458. If their fixed payment is $1,724, they paid down the principal by $266 ($1,724 - 1,458). Now they must pay 6.25% interest on the new principal balance of $279,734. This equals (279,734 x .0625) / 12, or $1,456.95.   >rates, payments 6<

37.   (b)   The second option, by 150.

The first option's interest total is (6.5% x $60,000) x 5 years, or $19,500. The second option will charge (6.25% x $60,000) x 5 years, plus $600, or a total of $19,350. The 2nd option is $150 cheaper.   >rates, payments 7<

38.   (b) $610

Carrie sold her home for (90% x $300,000), or $270,000. The loan she gave the buyer is for ($270,000 x 80%), or $216,000. The monthly payments on this are ($216,000 x 7%) / 12, or $1,260. Therefore she is making a profit of ($1,260 – 650), or $610. >rates, payments 8<

39.   (b) $43,990

First, Sally's projection of value is ($200,000) x105% x 105% x 105%, or $231,525. Second, the appraised value is ($231,525 x 90%), or $208,372. The loan amount on this @ 90% equals ($208,372 x .9) or $187,535.   The down payment required is ($231,525 - 187,535) or $43,990. >rates, payments 9<

40.   (b)   1.75 points

A discount point is one percent of the loan amount. Greg's loan is ($320,000 x 80%), or $256,000. If he paid $4,480, he paid 1.75% of the loan amount ($4,480 / 256,000), or 1.75 points. >points 1<

41.   (c) $3,375.

$225,000 x .015 = $3,375. Remember, one point = 1% of the loan amount.   >points 2<

42.   (c)   6.4%

On the loan with points, the lender will receive total interest of ($100,000 x 6%) x 5 years, plus ($100,000 x .02), or ($30,000 + 2,000), or $32,000. To receive this amount on a 5-year loan for $100,000, the lender will need to make ($32,000 / 5 years) or $6,400 / year. This return will require a 6.4% interest rate.   >points 3<

43.   (c) $15.

Without financing the points, the payments are ($200,000 x 5.75%) / 12, or $958. If the 1.5 points are added to principal, the new loan is for ($200,000 + 3,000), or $203,000. The new payment is ($203,000 x 5.75) / 12, or $973. Thus the payments increase $15. >points 4<

44.   (c) $150,000

Use the formula: Price x LTV Ratio = Loan. Then plug in the figures and calculate: Price x .80 = $120,000. Therefore, Price = $120,000 / .80 = $150,000. >LTV 1<

45.   (c)   No, she is short by $640.

The loan she can get amounts to ($240,000 x 80%), or $192,000. The points charge is ($192,000 x .02), or $3,840. Total closing costs are then $3,840 + 800, or $4640. Thus she is $640 short. >LTV 2<

46.   (c) $90,000

First, the sale price is 115% of the appraised value, so the appraised value is $230,000 / 115%, or $200,000. The lender will lend $140,000 (70% of appraised value), so the investor will have to come up with $90,000 ($230,000 – 140,000). >LTV 3<

47.   (c) $379,259

First, the annual interest paid is $1,600 x 12, or $19,200. The interest rate is 6.75%. Using the formula (Loan = Interest / Rate), the loan amount is $19,200 / 6.75%, or $284,444. As this is 75% of the price, the price is ($284,444 / .75), or $379,259. >LTV 4<

48.   (c) $181

The interest she pays in the first month is ($200,000 x 6.5%) / 12, or $1,083. If she pays $1,264 for PI, her principal portion of this payment is ($1,264 – 1,083), or $181. >LTV 5<

49.   (b) $840.

Monthly income qualification is derived by multiplying monthly income by the income ratio. Thus (36,000 / 12) x .28 = $840. Remember to first derive the monthly income. >loan qualification 1<

50.   (c) $1,040

The formula here is (housing debt + $600) / ($4,000 monthly income) = 41%. Solving for housing debt, we have (housing debt + $600) = (41% x $4,000). Thus (housing debt + $600) = $1,640. Subtracting out non-housing debt, we have (housing debt = $1640 – 600), or housing debt = $1,040. The borrower can afford $1,040 for PITI. >loan qualification 2 <

51.   (a) $810

The income ratio is equal to monthly housing expense / monthly gross income. To identify what the lender will allow, plug in the known and applicable numbers and solve: 27% = X / $3,000. (note the $500 does not apply to an income ratio; only to the debt ratio) Thus, 27% x $3,000 = $810.   >loan qualification 3<

52.   (a)   He can afford it by a $154 margin.

First, Thomas can afford to pay ($60,000 / 12) x 28%, or $1,400 / month. With $40,000 down, the loan amount must be $230,000. If the interest rate is 6.5%, the monthly payment will be ($230,000 x 6.5%) / 12, or $1246. Thus he can afford the property by a $154 margin. >loan qualification 4<

53.   (a) $206,000.

Remember the formula V = I / R where V is value, I is annual income, and R is the cap rate. Variations of this are: R = I / V in solving for the cap rate, and I = V x R in solving for income. Here, first identify net income by subtracting out vacancy and expenses. Then divide by the capitalization rate. Thus, ($30,000 – 1,500 – 10,000) / 9% = $205,555, or $206,000 rounded.   >cap rate, income value 1<

54.   (b) $400,000.

Value = Income / Cap rate. Thus, V= $20,000 / .05 = $400,000. >cap rate, income, value 2<

55.   (c) $15,000 overpriced.

Use the same formula V = I / R where V is the price and R is the rate of return. Then plug in the numbers to solve for V. The NOI of this property is ($60,000 - $22,000), or $38,000. The return is 11%. Therefore, the value to get this return must be $38,000 / .11, or $345,455. Since the price is $360,000, the price exceeds the amount needed for an 11% return by approximately $15,000 ($360,000 - $345,455 = $14,545).   >cap rate, income, value 3<

56.   (c)   5%.

Use the formula R = I / V. $25,000 / $500,000 = .05, or 5%. >cap rate, income, value 4<

57.   (b)   Yes, since he will yield 8.375%

Use the same formula R = I / V where V is the $2 million price, and R is the cap rate or rate of return. To identify income: (25,000 SF x $10/SF) = $250,000 gross income, minus 5% vacancy (.05 x $250,000), or $12,500, minus expenses of $70,000 = $167,500 net income. ($250,000 – 12,500 – 70,000) Now divide net income of $167,500 by $2,000,000 to derive the return of 8.375%. >cap rate, income, value 5<

58. (c) 50%

We must solve for income, so convert the V = I / R to I = V x R. We know V is the $1 million price, and R is the 10% cap rate or rate of return. Thus, the resort must net $100,000 ($1,000,000 x 10%.) Second, if expenses are 33% of gross income, then gross income must be $150,000 in order to net the $100,000. ($150,000 – 50,000 expenses = $100,000). Third, the total potential income is (200 nights x 10 rooms x $150 / night), or $300,000. So, to gross at least $150,000 from a potential of $300,000, the vacancy rate cannot exceed 50%.   >cap rate, income, value 6<

59. (d) $78,000.

Cost Approach formula: Land + (Cost of Improvements + Capital Additions – Depreciation) = Value. Thus you have $10,000 + ($75,000 - 7,000), or $78,000. >Cost approach 1<

60. (c) $475,000

Use the same Cost Approach formula: Land + (Cost of Improvements + Capital Additions – Depreciation) = Value. The land is worth (100,000 x 125%), or $125,000. Remember, you cannot depreciate the land, only the cost of the improvements. Therefore, annual depreciation is ($400,000 / 40), or $10,000. Total depreciation is ($10,000 x 5 years), or $50,000. Thus the value is ($125,000 + 400,000 – 50,000), or $475,000. >Cost approach 2<

61. (d) 150.

Use the formula: GRM = Price / Monthly Rent. Thus, $450,000 / $3,000 = 150. >GRMs 1<

62. (c) $203,000.

Use the formula: GIM = Price / Annual Income. To solve for price convert the formula to Price = GIM x Annual Income. Thus, ($1,200 x 12) equals $14,400 annual income. ($14,400 x 14.1 GIM) = $203,040, or $203,000 rounded.   >GRMs 2<

63. (c) 12.22

The GIM = Price / Annual Income. The annual income is ($15,000 x 12), or $180,000. Dividing this into the price of $2,200,000 derives a GIM of 12.22.   >GRMs 3<

64. (d) $1,778.

First calculate the total commission, then the co-brokerage splits, then the agent-broker split. Thus: $127,000 x 7% = $8,890 total commission. ($8,890 x 50%) = $4,445 total listing broker share. ($4,445 x 40% = $1,778 agent's share.   >commissions 1<

65. **(c)** **$4,278.**

The total commission is ($245,000 x 3%) = $7,350. The agent gets (65% x $7,350) - $500, which equals $4,278. >commissions 2<

66. **(b)** **$6,900.**

The total commission is 6% x $287,500, or $17,250. Since the agent sold her own listing, there is no co-brokerage split. Since she splits 60-40 with her broker, she will receive ($17,250 x .60), or $10,350 >commissions 3<

67. **(a)** **$9,669**

Figure the total commission, then the co-brokerage splits, then the broker-agent splits. Thus, ($425,000 x 7%) = $29,750. ($29,750 x 50%) = $14,875. ($14,875 x .65) = $9,669. >commissions 4<

68 **(a)** **$27,300**

The commission to be paid is ($210,000 x 7%), or $14,700, regardless of the co-brokerage split. Total closing costs will be ($14,700 commission + 3,000 closing costs + 165,000 mortgage balance), or $182,700. Subtracting this from the price nets Ryan ($210,000 – 182,700), or $27,300. >commissions 5<

69. **(a)** **7%**

First, the difference between the price and the seller's net was $50,000 ($620,000 – 570,000). Since $6,000 of the closing costs were non-commission costs, $44,000 is left for the commission. Regardless of the splits, the commission rate is ($44,000 / $620,000), or 7% >commissions 6<

70. **(a)** **6.5%**

First, if Graham receives $6,240, that is 60% of what the agency received. Thus the agency received ($6,240 / .60), or $10,400. If $10,400 is half the commission, the total commission is $20,800. This amount divided by the sale price produces the commission rate ($20,800 / 320,000), which equals 6.5%. The seller's other closing costs and net amount is unnecessary data. >commissions 7<

71. **(a)** **$2,547,619**

First, figure how much Hannah makes selling the $1 million. If the commission is 7%, her agency gets half, or 3.5%. Of this she gets half, or 1.75%. Thus, if she sells $1 million, she will earn $17,500 (1.75% x $1 million). Now she must earn another $32,500 under the new 60-40 arrangement. Here, Hannah makes 60% of 50% of 7%, or 2.1% (.07 x .6 x .5). In order to make $32,500 @2.1%, Hannah must sell ($32,500 / .021), or $1,547,619. Thus her total volume must be $2,547,619. >commissions 8<

**72.	(d)	Plan B is better by $14,250.**

First, William's sales goal is $5 million ($4 MM x 125%). At the 65-35 rate, his average commission will be (7% x 50% x 65%), or 2.275%. Thus he will make $113,750 on this plan. On the other plan, his average commission will be (7% x 50% x 80%), or 2.8%, but he must also pay $12,000 in desk fees. Thus he will make ($5 MM x 2.8%) - $12,000, or ($140,000 – 12,000), or $128,000. This is the better plan by $14,250. >commissions 9<

**73.	(d)	$47,640**

On the two homes, she made ($160,000 price x 2 homes x 6% commission x 50% co-broke x 60% agent share) or $5,760. On her 3 listings she made ($640,000 sales x 7% x 60% share), or $26,880. On the land, she made ($250,000 price x 10% x 60% share), or $15,000. The sum of these amounts is $47,640. >commissions 10<

**74.	(a)	A debit to the buyer and credit to the seller for $375.**

Here, the seller has paid for 25% of his part of the premium, and 75% of the buyer's part. Therefore, the buyer must be charged, or debited for 75%, or $375, and the seller credited by the same amount. >prorations 1<

**75.	(d)	A debit to the seller and credit to the buyer for $680.**

Since the seller received a portion of the buyer's income, the seller must be debited ($680) for the buyer's share, and the buyer credited ($680). > prorations 2<

**76.	(b)	A debit to the seller and credit to the buyer for $100.**

Here, the seller has incurred $100 of the $300 expense that the buyer will be paying for after closing. Thus the seller must pay the buyer the $100 portion of the expense he or she used. Therefore, debit the seller and credit the buyer $100. > prorations 3<

**77.	(c)	A credit to the buyer and debit to the seller for $785.34.**

The daily tax expense, first, is ($3,150 / 365) or $8.63. Since the buyer will pay the taxes after closing, the seller must pay the buyer his or her portion of the tax bill at closing, which is the 91 days from the beginning of the year through closing. Therefore, credit the buyer and debit the seller ($8.63 x 91), or $785.34. > prorations 4<

**78.	(b)	A debit to the buyer and credit to the seller for $183.33.**

This method assumes all months are 30 days and the year is 360 days. The daily proration is therefore $375 / 360, or $1.04. The closing occurs on the 184th day of the year. Thus the seller's share is ($1.04 x 184) = $191.67. The buyer's share is ($375 – 191.67), or $183.33. Charge, or debit the buyer and credit the seller $183.33. Note that if the day of closing is the buyer's, there would be 183 day's worth of prorated expense. > prorations 5<

79.   (d) $735.34.

Assuming a 365 day year, the daily tax expense is ($1,100 / 365), or $3.013. As taxes are paid in arrears, the buyer will be paying the annual bill. Thus, he will be owed a credit for the seller's share of the bill, which is $3.013 x 244 days, or $735.34.   > prorations 6<

80.   (a) $507.50.

If the buyer pays $525 interest for 30 days, the daily expense is ($525 / 30), or $17.50. If there are 29 days of pre-paid expense, the buyer's charge is ($17.50 x 29), or $507.50. >prorations 7<

81.   (d)   Credit buyer $368.00.

For the monthly proration using the 365-day method, solve first for the daily rent amount: ($713 / 31), or $23. Since the landlord received the rent and owes the buyer portions of the rent, the buyer will be credited. The owed amount is for the 16th through the 31st, or 16 days, since the closing day belongs to the buyer. Therefore, credit the buyer and debit the seller ($23 x 16),or $368. >prorations 8<

82.   (d) $86,372

First calculate the transfer tax: ($322,600 / 500) = 645.2 units of $500. Round this up to 646, then multiply times $1.00 to get $646 transfer tax cost. Next figure the commission @ ($322,600 x .07), or $22,582. Next, the seller's real estate tax proration charge will be $2,000. Then, add up the expenses: ($646 transfer tax + 450 title + 550 attorney + 22,582 commission + 210,000 loan payoff + 2,000 tax proration) = $236,228. Subtracting this from the sale price = $86,372. >net proceeds 1<

83.   (d) $57,854

First, the credits applied to the price are the deposit ($5,000) and the loan ($292,000 x 80%), or $233,600. This leaves him owing ($292,000 – 5,000 – 233,600), or $53,400. Henry's closing expenses are $3,504 for points ($233,600 x .015), $1,200 for fees, a tax credit of $900 ($3,600 x 25%), and $650 for homeowners insurance. His cash required at closing is thus ($53,400 + 3,504 + 1,200 – 900 + 650), or $57,854.   >net proceeds 2<

84.   (d) $5,848.

Be careful here. Since the net price is $5,000, the taking price (TP) minus the commission must equal the net price. In other words, the net price is 90% of the taking price. Since TP x 90% = Net, TP = Net / 90%. So the taking price is $5,556 ($5,000 / .9). Apply the same logic to deriving the asking price: the taking price is 95% of the list price, therefore the list price = (taking price / 95%), or $5,848. Now work backwards to prove your answer: (5,848 x 95% margin x .90 net of commission) = $5,000 >net proceeds 3<

85.   (b) $237,234

Use the formula (net + closing costs + loan payoff) / (100% - commission rate). Thus, ($70,000 + 3,000 + 150,000) / .94 = $237,234. Working backwards to prove this, you have $237,234 price – 14,234 commission – 3,000 closing costs – 150,000 loan payoff) = $70,000. >net proceeds 4<

86.   (b) $55,000

First figure what her net must be to generate a 20% profit: The profit is ($40,000 x 20%), or $8,000. The net must then be $48,000. Now to find the sale price, use the formula (net + closing costs + loan payoff) / (100% - commission rate). Thus, ($48,000 + 1,500) / .90 = $55,000. Working backwards to prove this, you have ($55,000 price – 5,500 commission – 1,500 closing costs) = $48,000 >net proceeds 5<

87.   (c) $96,000.

Since the comparable has an extra half-bath, it is adjusted downward to equalize with the subject. Conversely, since it has no patio, the appraiser adds value to the comparable. Thus, $100,000 minus $5,000 plus $1,000 equals $96,000.   >Adjusting Comps 1<

88.   (c) $339,500.

Adjust the $320,000 comparable as follows: -5,000 for the bath; + 15,000 for the pool ; + 13,000 for the lot; and - 3,500 for the porch. Thus the value of the subject is ($320,000 – 5,000 + 15,000 + 13,000 – 3,500), or $339,500. >Adjusting Comps 2<

89.   (a) $2,100.

Annual gross operating income ($1,500 x 12 = $18,000) - annual operating expenses ($500 x 12 = $6,000) = annual net operating income ($12,000); annual net operating income ($12,000) - cost recovery expense ($5,000) = taxable income ($7,000); taxable income ($7,000) x tax rate (30%) = tax liability ($2,100). >tax liability 1<

90.   (c) $2,240

The basic formula for tax liability is: Taxable Income x Tax Rate = Tax Liability. Taxable Income is Net Operating Income - Interest Expense - Cost Recovery Expense. Therefore, the annual tax is $50,000 (NOI) - $35,000 (Interest Expense) - $7,000 (Cost Recovery Expense) x 28% = $2,240. Note that the principal payment is not deductible in calculating taxable income.   > tax liability 2<

91. (c) $508

The total assessed value is ($84,550 + 235,000), or $319,550. The annual tax is based on ($319,550 / 100) = 3195.5 100's. Round up to 3196. To derive the annual tax, multiply 3,196 x 1.91, or $6,104.36. Divide this by 12 for the monthly escrow: ($6,104 / 12) = $508.   > taxation 6<

92. (d) 35%

The formula for profit % is (profit / initial investment). The profit made was ($18,000 x 3) − 40,000 initial investment, or $14,000. Dividing this by the amount invested derives a profit percent of 35%. > cap gain and basis 5<

93. (a) 22,787

First, convert the area into acres: (174,240 / 43,560 sf per acre) = 4 acres. With the loss factor, Charlene has 3 acres to convert into ¼ acre lots, so she can yield 12 lots total. Her land cost is (174,240 x $1.25), or $217,800. Thus, her cost /lot is ($217,800 / 12), or $18,150. Her profit must be 25% of this, or ($18,150 x .25), or $4,537.50. So the price must be ($18,150 + 4,537.50), or $22,787.50 > cap gain and basis 6<

94. (d) $144,960

First, the loan amount is ($320,000 x 90%), or $288,000. Monthly PI payments on the 30-year loan are ((loan amount / 1000) x loan constant, or ($288,000 / 1000 x 6.3207), or $1,820. Total payments are ($1,820 x 360 months), or $655,200. Total interest paid equals ($655,200 - $288,000 loan principal), or $367,200. Monthly PI payments on the 20-year loan are ($288,000 / 1000 x 7.3824), or $2126. Total payments are ($2126 x 240 months), or $510,240. Total interest paid equals ($510,240 - $288,000 loan principal), or $222,240. The 30-year loan therefore costs $144,960 more. > interest rates, payment, term 10<

95. (d) $92,000

The formula for equity is (current value – indebtedness). The current value is ($180,000 + 25% x 180,000), or ($180,000 x 125%), or $225,000. The current debt is ($180,000 x 80%) - $11,000, or $133,000. Their equity is therefore $92,000. > equity 1<

96. (d) $197, 352

First, the Poormons can afford a monthly PITI payment of ($82,000 / 12) x .28, or $1,913. Now take out monthly taxes and insurance to derive their maximum PI payment: ($1,913 − 600), or $1,313. ($6,000 annual taxes and insurance = 500 /month). Now use the formula: Loan amount / 1000 x constant = monthly payment. Thus, (Loan amount) / 1,000 x 6.6531 = $1,313. Solving this we have Loan amount = 1,313 x 1,000 / 6.6531, or $197, 352.   > loan qualification 5<

97.   (a) $30,900

Their fixed rent is (1,500 SF x $1.25/SF) x 12 months, or $22,500. The percentage rent is ($35,000 x .02) x 12, or $8,400. Total rent is ($22,500 + 8,400), or $30,900. > percentage leases 1<

98.   (b)   42,550 SF

The total lot measures (250' x 200'), or 50,000 SF. Of this, take off the Broadway setback of 15': (15' x 250', or 3,750 SF. This leaves a depth measurement of 185' on both sides. Now, take out the setback area on both sides of the lot: (185' x 10') x 2 = 3,700 SF. The net buildable area is now (50,000 SF – 3,750 – 3,700, or 42,550. > area measurements 11<

99.   (b) $188,000

First, Bernie's lot @.75 acre totals 32,670 SF (43,560 x .75). The front footage is (32,670 / 250' depth), or 130.68'. Market value @ $1,250/front foot equals ($1,250 x 130.68), or $163,350. Add 15% to this number and round off to get the listing price: ($163,350 x 115%) = $187,852, or $188,000. > area measurements 12<

100.   (b)   Debit seller, credit buyer $866.70

The rent received equals ($650 x 4), or $2,600. There are 20 seller days and 10 buyer days. The daily proration is $2,600 / 30, or $86.67. Here the seller received the funds therefore he owes the buyer. Debit seller and credit buyer the buyer's portion: ($86.67 x 10 days), or $866.70. >prorations 13<

101.   (b) $166,150

Use the formula: Gain = (Net selling price – adjusted basis) where adjusted basis = (beginning cost – depreciation). The selling price is $240,000 x 5% annual appreciation x 10 years, or ($240,000 + 50% x 240,000), or $360,000. Since land cannot be depreciated, the depreciable basis is ($240,000 total cost – 60,000 land value), or $180,000 (land = 25% total value). Annual depreciation = ($180,000 / 39 years), or $4,615. Thus total depreciation is ($4,615 x 10 years), or $46,150. The adjusted basis is therefore ($240,000 – 46,150), or $193,850. The total gain is therefore ($360,000 – 193,850), or $166,150. >cap gain, basis 5<

102.   (a)   7.5%

First, if she borrows $8,000, she is paying $900 interest over 1.5 years. Annual interest paid is therefore ($900 / 1.5), or $600. Using the formula (Rate = Interest / Principal), her interest rate is $600 / $8,000, or 7.5%. >rate, payment, term 10<

103.   (d) $187,500

Use the formula: (Percent of insurable property value carried / 80% replacement cost) x claim = recovery, where the insurable property value variable excludes the land value and is valued at replacement cost. Here, the insurable portion of the property is ($320,000 - 25% land value), or $240,000. The Jacksons are carrying insurance to cover 75% of the replacement cost of the entire property. Their recovery amount is therefore (75% / 80%) x $200,000, or $187,500. >insurance 1<

104.   (c) $140,000

Use the formula: (Percent of insurable property value carried / 80% replacement cost) x claim = recovery. Thus, (75% / 80% x $150,000) = $140,625. However, the face value of the policy is the maximum they can receive, which is $140,000 >insurance 2<

105.   (a)   a debit of $250 to the seller.

Since the entire tax bill is due and payable by the seller, Skip will be debited the entire amount of the tax bill and the bill will be paid. <prorations>

106.   (a)   Debit seller and credit buyer $200,000.

Mortgage assumptions are effectively inflows of money, hence a credit, to the buyer and purchase price money the seller does not receive at closing, hence a reduction or debit of the seller's proceeds. <prorations>

107.   (c)   A debit of $190,000 to the sellers.

The loan payoff is money paid to the lender from the purchase price funds, therefore a debit to the seller's proceeds. The debit amount is the loan balance, not the equity amount. <prorations>

108.   (c) $80,000

Applying the formula (percent of insurable property value carried / 80% replacement cost) x claim = recovery), divide the amount of coverage carried ($160,000) by 80% of the insurable property value ($250,000) to get the percent of the claim the company will pay (80%). Multiply this percentage by the claim amount to get $80,000, what the company will pay.   <insurance>

109.   (d) $432.90

First, the loan amount is 75% of $400,000, or $300,000. Per the formula, first divide the loan amount by 1,000 to get 300. Multiply this times the loan constant to derive the monthly PI payments: 300 x 6.443 = $1932.90. Now calculate the first month's interest: $300,000 x 6% / 12, or $1,500. The principal payment is the difference: $1932.90 – $1500 = $432.90 <finance>

110. (a) 70%

Here you must subtract beginning equity from ending equity to identify the equity gain. Then this is divided by the initial equity to figure the equity gain as a percent. Then divide by five to get the average annual equity gain percent. Thus, first, the initial loan amount is $220,000 x 80%, or $176,000. Thus her initial equity is ($220,000 value - $176,000 loan), or $44,000. After five years, her loan balance is ($176,000 – 14,000), or $162,000. Her ending equity is therefore ($360,000 value – $162,000 loan balance), or $198,000. Her total equity gain is thus $198,000 – 44,000, or $154,000. As a percent, her total equity gain is $154,000 / $44,000 initial equity, or 350%. Averaging this per year, we get 350% / 5 years , or 70%. <investment>

111. (b) Betty qualifies under the income ratio, but not the debt ratio.

Using the income ratio to qualify, ($5,600 x 28%) = $1,568. Since this exceeds $1289, she qualifies. Under the debt ratio, she can spend up to (36% x $5,600), or $2,016. Betty's total debt payments would be ($1,289 + 400 + 525), or $2,214. This exceeds $2,016 so she does not qualify under the debt ratio. <finance>

112. (b) 40 acres

First, remember that a section contains 640 acres. The area in question is a forth of a half of a half of the total section. So divide 640 by (4 x 2 x 2). 4 x 2 x 2 is 64 and 640/10= 40 acres. <legal description areas>

113. (c) 10 acres

This key to this question is to recall that a section has 640 acres. Theresa's property only is a ¼ of a ¼ of a ¼ of the entire section. Multiply 4 x 4 x 4 which equals 64. Theresa owns 1/64 of the section. Divide 640 by 64 and you get the answer of 10 acres. <legal description areas>

114. (c) 92%

To find the percent of listing price the offer is, divide the offer by the listing price. In this question the offer is $689,000 and the listing price is $749,000. $\frac{\$689,000}{\$749,000} = 92\%$

115. (a) $10,920

In this question, ignore the appraisal value (it is a distractor). To find the earnest money amount, multiply the offer ($312,000) by the percent required by the seller (3.5%). $312,000 \times .035 = \$10,920$

Order Form

If you liked Real Estate Math Express, check out the other titles of Performance Programs Company!

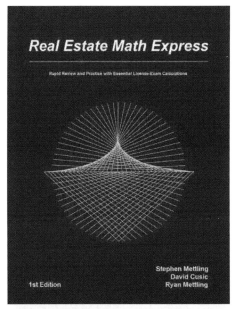

Where can you buy Real Estate Math Express?

Real Estate Math Express is available through nearly all online retailers.

Looking for a real estate principles textbook? Use Principles of Real Estate Practice!

If you are currently enrolled in a real estate licensure course, you need Principles of Real Estate Practice to pass your real estate licensure test! It contains the essentials of real estate law, principles, and practices taught in real estate schools and colleges across the country, including all those fundamentals that real estate educators, practicing professionals, national testing services, and state licensing officials agree are necessary for basic competence. For several states, we now have state-specific versions.

Where can you buy Principles of Real Estate Practice?

Copies of Principles of Real Estate Practice and its state-specific versions are available through online retailers.

Need to prepare for your state's real estate exam? Prepare with the Real Estate License Exam Prep (RELEP) series!

Performance Programs Company has launched its new line of state-specific exam prep titles. Each title has comprehensive national and state-specific key point reviews, over 500 national questions, and 150 state questions. Copies are available through online retailers.

Publisher Contact

Ryan Mettling
Performance Programs Company
502 S. Fremont Ave., Ste. 724
Tampa, FL 33606
ryan@performanceprogramscompany.com
www.performanceprogramscompany.com

75

Made in the USA
Lexington, KY
12 December 2019